Core Issues

Dissecting Nuclear Power Today

Steve Kidd

Special **NUCLEAR ENGINEERING INTERNATIONAL** *Publications*

Nuclear Engineering International Special Publications

All rights reserved. No part of this publication may be reproduced or transmitted in any form or by any means, including photocopying and recording, without the written permission of the copyright holder, application for which should be addressed to the Publishers. Such written permission must also be obtained before any part of this publication is stored in a retrieval system of any nature.

ISBN 978-1-903077-56-6

© 2008 Progressive Media Markets Ltd

Typesetting by Karen Townsend
Cover design by Natalie Kyne
Printed and bound by Polestar Wheatons, Hennock Road,
Marsh Barton, Exeter, Devon EX2 8RP, United Kingdom

Nuclear Engineering International Special Publications, Progressive House,
2 Maidstone Road, Foots Cray, Sidcup, Kent DA14 5HZ, United Kingdom
www.neimagazine.com

About the Author

Steve Kidd is Director of Strategy & Research at the World Nuclear Association (WNA), the international association for nuclear energy based in London.
After reading economics at Queens' College Cambridge and a brief period teaching and researching at Sheffield University, he followed a career as an industrial economist with leading UK companies. He practised as an independent consultant from 1990 and then joined the former Uranium Institute as Senior Research Officer in 1995. He assumed his present position when the Institute changed its name to the World Nuclear Association in 2001.

He acts as Secretary to many WNA working groups, notably those preparing incisive reports such as the biennial treatise on the global nuclear fuel market. He authors many articles on the commercial aspects of nuclear and is a frequent speaker at conferences and meetings around the world, particularly those targeted at non-specialists. Finally, he organises and teaches at training courses for nuclear professionals in developing nuclear countries, on behalf of the World Nuclear University (WNU).

For my parents, Don and Joan Kidd, without whom...

Contents

Preface   xi

1. Nuclear around the world   1

explains how the experience of nuclear power (and its prospects) differ considerably from country to country. Nuclear power now accounts for almost 16% of world electricity generation, but the last 15 years have been characterised by relative stagnation. Now there are renewed prospects for further reactors in both established nuclear countries and some new ones. The major developing countries, China and India, are particularly important. Yet there remain significant anti-nuclear forces in several European countries, notably Germany.

Introduction	1
China: the major growth centre?	2
India: tomorrow's giant?	7
Argentina and Brazil: a Latin revival?	10
Japan: where to now?	14
South Korea: a success story?	18
Germany: is it really facing a phaseout?	21
Eastern Europe: a lively bright spot?	25
Russia: where's it heading?	29
New nuclear countries: where and when?	33

2. Nuclear new build and economics   37

highlights that, despite the increased attention now being paid to nuclear for environmental and security of energy supply reasons, it's clear that new nuclear plants must prove their economic credentials in the power markets of today. If it can be shown that a new nuclear plant will provide the cheapest baseload electricity over the long term, this is a very powerful argument in favour of selecting it. Several possible constraints have been mentioned, notably the availability of finance in liberalised markets and lack of supply capability, but these can be overcome.

Introduction	37
New nuclear plants: can we ever again build in volume?	39
New nuclear build: do the economics stack up?	42
Liberalised power markets: how can nuclear fit?	46

Financing: a barrier to new nuclear build?	50
New nuclear build: is there sufficient supply capability?	55
New nuclear plants: what are the real issues?	58

3. Public acceptance 63

shows that the roots of opposition to anything to do with nuclear run very deep and are an essential element of the environmental movement, which has found the industry a relatively easy target. Experience has shown that gaining public approval is best accomplished at the local level, by letting people visit nuclear facilities and ask lots of questions. For the industry, the best approach is to carry on doing what it does best, which is running the facilities well, both safely and economically.

Introduction	63
Nuclear's adverse public image: where are the roots?	65
Nuclear's low external costs: justifying public support?	68
Potential weapons proliferation: a real risk?	72
Nuclear and renewables: can they be partners?	76
Public opinion: why hasn't it yet been turned around?	80
The serious opposition to nuclear: how can it be handled?	84
The battle to win hearts: how can it be won?	87

4. Nuclear fuel 93

explains that just because the low cost of nuclear fuel and its relative stability over time has always constituted a prime economic advantage of nuclear power, this doesn't mean that the fuel sector is unimportant or lacking in interest. Indeed, the opposite is very much the case, partly because the front end of fuel cycle is, in itself, quite complex, with individual markets for each of uranium, conversion, enrichment and fuel fabrication. In particular, the upsurge in world uranium prices since 2003 and the subsequent sharp fall back from the peaks have in themselves created a lot of comment and interest (particularly from the financial sector). This has stimulated a renewed look at alternatives to the traditional ways of buying and selling nuclear fuel and encouraged buyers to press for the opening up of as many sources of supply as possible.

Introduction	93
Nuclear fuel demand: will it carry on rising?	95
WNA Market Report: calling the shots?	99

Uranium: is there sufficient to fuel nuclear growth?	103
More uranium: when and from where?	106
Uranium: more production needed post 2013?	110
Junior uranium companies: at last something new?	114
Enriching: is it more than a proliferation risk?	117
Additional enrichment: a substitute for expensive uranium?	121
Enrichment: as interesting as uranium?	125
Fuel fabrication: a different sort of market?	128
The uranium boom: was it predictable?	132
Uranium prices: providing appropriate signals?	135
Uranium market: what needs to happen?	138
Nuclear fuel: some positive changes?	142

5. **Trade and the back end of the fuel cycle** 147

   discusses the significant international restrictions on knowledge transfer and trade that are important in nuclear commerce and also important issues in the management of used fuel and the decommissioning of nuclear facilities. The constraints are very much bound up within the provisions of the Treaty on the Non-Proliferation of Nuclear Weapons (NPT) and its policing by the Nuclear Suppliers Group (NSG). What to do with the used fuel when it is discharged from a nuclear reactor has been one of the biggest issues since the early days of nuclear power and has given rise to probably the biggest issues on public acceptance. The debate over reprocessing is an important part of this. Resolution of these important issues is vital to the future of nuclear as is proof that the industry will take full responsibility for returning closed nuclear sites to alternative use.

Introduction	147
Nuclear commerce: do restrictions hinder future growth?	148
Nuclear's Achilles' heal: or own goal?	152
Nuclear waste: what's happening now?	156
Reprocessing used fuel: devil or saint?	160
Recycling uranium and plutonium: where's it heading?	164
Decommissioning activity: the sunset for nuclear?	168

6. **The big picture** 173

puts nuclear within the context of wider world energy developments. These have now returned as a subject of popular debate, after many years on the sidelines following the world oil crises of the 1970s. This has been prompted by renewed concerns over the security of long-term oil and gas supplies (indicated by significant price escalation) but also by the concerns about the environmental consequences of continued mass exploitation of fossil fuel resources. There are various scenarios now postulated for the future of nuclear power but lifecycle analysis of various electricity generation options shows nuclear in a very favourable light. Yet nuclear proliferation remains as a very live issue and could conceivably threaten prospects of a nuclear revival. Amongst other issues, this is addressed by initiatives such as the Global Nuclear Energy Partnership (GNEP) proposed by the United States and alternative (largely complementary) suggestions by other countries.

Introduction	173
World energy in the 21st Century: how much nuclear?	174
IEA World Energy Outlook: where is nuclear?	178
The nuclear industry: how strong is it?	182
Nuclear: an odd or normal business?	186
Nuclear: is there any net energy addition?	189
Nuclear proliferation and the terrorist threat: a barrier to new build?	193
GNEP: the right way forward?	197
Changing international arrangements: what are the new initiatives?	202

Summing up 207

Glossary 211

Preface

The foundations for this book are the monthly *Comment* articles I have been writing for the magazine *Nuclear Engineering International* for the past four years. When the Editor, Stephen Tarlton, asked me to contribute on a monthly basis, I don't think either of us had much idea how long it would be before I ran out of ideas or, alternatively, general boredom overwhelmed the two of us. Yet as things have transpired, I have found writing the columns relatively easy and also very enjoyable, while the general feedback from readers has been positive. We therefore decided to put together an edited version of my first 50 articles, in the hope that some of my enthusiasm for energy and specifically nuclear matters would transfer to even more readers.

Although those new to nuclear power can certainly tackle this book, it may be particularly appropriate for those who have some grounding in one or more of the aspects of the wide nuclear area, but seek wider thoughtful enlightenment. For the absolute beginner, I would wholeheartedly recommend my colleague Ian Hore-Lacy's primer for the World Nuclear University, *Nuclear Energy in the 21st Century*.

Core Issues is set out under six major chapter headings, each covering an important grouping of issues, with a final, relatively brief, summing up. Although there is some logical progression within both the chapters and the overall book, because the original 1500 word articles form the grounding, it is possible to jump around rather than read from cover to cover. Indeed, I encourage the reader to do so.

The gestation period for the book has been an exciting one for nuclear power, marked by an explosion of public interest. This has been caused by a subtle mixture of environmental, energy security and economic developments, favourable to a new round of reactor construction. The industry is, however, very complex and not so easy to explain to the outsider. Many texts concentrate on the scientific and technical side, but as an economist, my interest is more in the commercial aspects. Much of my work therefore centres around analysing and then explaining how the various nuts and bolts of the industry, from uranium exploration to plant decommissioning, fit together in a cohesive whole. My hope is that this book will contribute to this.

I would like to acknowledge the contribution of my colleagues at the World Nuclear Association to the development of my thinking about the industry and also that of many representatives from our member companies around the world. Particular thanks must, of course, go to Stephen Tarlton for his ideas and editorial skills and also to his team at the magazine. Finally, my family and friends must be thanked for their patience when I've switched off, deep in thought and not acknowledging their concerns as carefully as I should have been. But all errors and omissions in the book are my own.

Steve Kidd
April 2008

1. NUCLEAR AROUND THE WORLD

Introduction Nuclear power now accounts for almost 16% of world electricity generation, with 439 reactors in operation (end of 2007) in 30 countries. After a period of rapid growth in the late 1970s and 1980s (over 200 reactors came into operation during the 1980s alone), the last 15 years have been characterised by much slower growth and, indeed, relative stagnation (if reactor startups are considered – in recent years, only 3-5 new reactors have been starting up annually). The reasons for the slowdown are a mixture of economics and difficulties with public acceptance. Slower electricity demand growth and new plants taking too long to build (and performing badly once in operation) are both important, as are relatively lower fossil fuel prices and the introduction of innovative alternative generating technology such as combined cycle gas turbines (CCGTs). The accidents at Three Mile Island in 1979 and at Chernobyl in 1986 added to public concerns over safety and radiation exposures, while the inability to move forward quickly with facilities for the disposal of radioactive waste damaged the image of the industry.

Nevertheless, the nuclear share of world electricity at 16% has remained almost constant since the late 1980s, despite the significant slowdown in new reactor startups. The explanation for this is that the reactors in operation have gradually been performing at increased capacity factors (essentially the percentage of time they are online), while reactors have also received extensions to their licensed lives and uprates to their rated power outputs. From an atmosphere of 'doom and gloom' in the industry in the late 1990s, with a widespread expectation of a slow and lingering death as reactors gradually shut down and were decommissioned, the improved operating performance has constituted the foundation of what has been called a 'nuclear renaissance'. The excellent safety record since Chernobyl combined with increased interest in nuclear owing to its environmental and security of supply credentials have added to this. There is now a widespread expectation that most of today's operating reactors will stay in service for many years in the future, in many cases for far longer than was envisaged when they were started up. In addition, there is now a belief that many new reactors may enter operation in the future, with increasing numbers of people seeing nuclear as an important part of satisfying 21st Century energy requirements in an economic and environmentally sustainable way.

The experience of nuclear power (and its prospects) differs considerably around the world and this chapter considers some of the more interesting countries and regions. In both China and India, nuclear currently only accounts for a small part of electricity generation, but the prospects for new plants are exciting in two huge and rapidly-growing economies. Their programmes could well dominate the immediate future, before new build programmes get re-established in some of the early pioneers in nuclear, such as the United States and United Kingdom. In Latin America, nuclear established early roots in both Argentina and Brazil but expansion got cut off at an early stage, as the world industry went into relative decline. There are now, however, good prospects for a return to the previous growth paths. Japan and South Korea, by contrast, have much more mature nuclear programmes, and are both at the stage of adding further reactors to maintain their nuclear shares at high levels. They are also now capable of exporting their own nuclear technology to other countries.

Within Europe, there are significant differences in attachment to nuclear. In France, nearly 80% of electricity generation is from nuclear, while in others such as Italy it's zero. Germany is a particularly interesting case as it is a leader in nuclear technology, with reactors operating very safely and economically, but which has a nuclear phaseout enshrined in law. This clearly offers a challenge to the nuclear renaissance and what happens there is significant for prospects everywhere else. By contrast, in Eastern Europe, where many countries have current reactors, nuclear has some good expansion prospects. Russia has been rather a 'sleeping giant' in the commercial nuclear sector since the fall of the Soviet Union, but has great knowledge and experience of nuclear technology. This is at last now leading to expectation of a significant new build programme by 2020. Finally, there are now a large number of countries expressing an interest in nuclear, with hopes of reactors starting up in many new countries by 2030. It is sometimes difficult to disentangle all the various media announcements in this area along with the lack of realism in timings of programmes, but it is clear that there are reasonable expectations in several countries by 2020-2025.

China: the major growth centre?

The media is full of dramatic growth stories about Chinese economic development, as if it began only yesterday. In fact, China has been growing very quickly for almost two decades now, but the low base in 1980 meant that even a country of 1.4 billion people has only now begun to have an appreciable influence in the world economy. The low-cost manufacturing base stands out as formidable to overseas competitors, while the demands for energy and raw material imports are having an important influence on world markets. But what about China's actual and potential influence in the world nuclear industry?

China is one of the five nuclear weapons states and its nuclear sector still bears some hallmarks of its previous domination by military objectives. In fact, it bears some resemblance to the Russian industry ten years ago, as there is some secrecy about facts and figures combined with slowly emerging understanding of the requirements of a commercial nuclear sector. For example, there are no official figures for Chinese uranium production, while the need to have all plans closely scrutinised and approved by central government at least threatens to hold nuclear back. Yet China, unlike India, is not constrained by an inability to import technology and materials from abroad – indeed, China is a member of the Nuclear Suppliers Group (NSG), which forbids trade with India owing to its refusal to sign the Treaty on the Non-Proliferation of Nuclear Weapons (NPT).

China currently has 11 nuclear reactors in operation, with two others about 50% complete and a larger group at a very early stage of construction. Four are of Areva NP origin, two are AECL CANDUs, a further two are Russian VVER-1000s with the other three of Chinese origin. The combined capacity of the 11 reactors is 8.6 GWe.

To put this into context, it is really a drop in the ocean within Chinese electricity supply, accounting for only 2% of installed capacity. In recent years, China has been commissioning over 60 GWe of electricity generating capacity annually, predominantly coal-based but also including a lot of hydro. Indeed, the Three Gorges Dam will itself add 18GWe to capacity, double nuclear's total (yet the prime motivation for this monumental project was more to do with the need for flood control than power).

The interesting question now concerns the likely extent of the future nuclear construction programme and in particular how significant it will be in world terms. It is clear that nuclear provides no immediate solution to China's growing power requirements, which in some cases have resulted in blackouts for industrial plants in the warm summer months. This must be sought by demand control (including a move to more realistic market-based pricing) and further additions to the stock of fossil fuel generating plants. Over the longer run, however, it is possible to see nuclear increasing its share of total electricity generation, with nuclear capacity rising to around 40 GWe by 2020, which should be 4-5% of the total by then. Indeed, this figure has been sanctioned in official circles as a target. To get there, however, is going to require a nuclear programme of a similar magnitude to that of France in the 1980s, with two or three 1000 MWe reactors coming into operation each year from 2010-2020. Is this realistic?

Gaining governmental approval for new reactors has certainly been a torturous process, but now appears to have eased up considerably. In the 9th (1996-2000) and 10th (2001-2005) Economic Plans, nuclear fared comparatively badly. This has been blamed on concerns about the economics of nuclear compared with alternative technologies and also a desire to spur economic development away from the prosperous coastal areas (where nuclear plants would be located) in favour of the under-developed central and western areas of China (where hydro and coal are centred). Delays to approving reactor plans ensured that after the twin VVER-1000s came into operation in 2006-7, there would be a gap of a few years before anything else starts up, as construction is only well-advanced on two further Areva-origin 1000 MWe pressurised water reactors (PWRs) – now with high local Chinese content and referred to as CPR-1000 units – at Ling Ao in Guangdong province. Much attention has been placed on the Chinese adoption of 'Generation III' nuclear technology, by signing agreements with Western vendors to introduce their latest designs in China. This has now been resolved with the signing of a technology transfer agreement for four Westinghouse AP1000s to be built at two new sites in China and an additional deal to build two Areva EPRs at another new location in Guangdong province. There is also a plan to build a further two VVER-1000s at the Tianwan site. Beyond this, there are now two further Chinese design reactors under construction at the Qinshan site, while the China Guangdong Group has an ambitious schedule to build many more CPR-1000 reactors at various sites such as Hongyanhe and Ningde.

It now appears that the nuclear plans mentioned in the 11th Five Year Plan (2006-2010) have all gained the necessary approvals to allow 40 GWe by 2020 to be achievable. The environmental advantages of nuclear have been noted by state planners and political leaders, with latest national plans containing detailed environmental targets on energy use and carbon emissions. The emissions of coal-fired plants near the crowded cities have been recognised as a particular problem. It has also been appreciated that the existing reactors have come into operation on or even ahead of schedule and also on cost. They are also running at high load factors and the safety record is good.

The contribution that a nuclear power plant makes to economic development in local areas has also been well-noted. Indeed, this spurring of local economic development has led to many provincial governments in China making preliminary plans for nuclear power plants in their areas and not just in the coastal provinces. Outside commentators often make the mistake of ignoring regional political power in China, wrongly seeing it as a monolithic centralised and undemocratic state. Yet within one-party rule by the Communists, the provinces have a lot of sway and there are routes for local opinion to pass through into policy in Beijing, through various consultative committees. There is therefore an unusual situation in China, whereby it appears to be the local people who are pressing the government for more nuclear power plants, rather than the government sweeping the issue under the carpet (largely for fear of public opinion about nuclear, as is the case in most of the rest of the world). There is, however, some concern within the Chinese industry about the future of public opinion on nuclear, particularly as the country gets richer. Yet there is determination to learn lessons from the mishandling of key issues by the Western industry, such as safety, waste management and proliferation.

There are therefore good grounds to see 40 GWe by 2020 as realistic. Unless there is a strong revival of nuclear in North America and Western Europe, this is likely to be the main growth area for the nuclear industry over this period (notwithstanding similar growth plans, but on a smaller scale, in India). The main issue is now the ability to complete so many new reactors of different designs on various sites throughout the country. A huge training programme is underway to develop the necessary specialist staff but the sheer magnitude of economic activity in China today means that nuclear cannot have the degree of national priority it had in France. Against that, China is far bigger than France and has more powerful regional bases, which can also drive the programme in addition to the national government.

There is currently a keen debate within China about the degree of reactor design standardisation that should take place. With long-term aims expressed for nuclear to reach 120-160 GWe by 2030, it should be possible to build more than one reactor design in sufficient numbers to still achieve significant economies of scale. It is clearly the intention to evaluate the foreign Generation III technologies with a view eventually to follow the Korean route of developing and then adopting one or more standard Chinese nuclear plants, which will be built in great numbers around the country. This is likely to be a PWR hybrid of existing Chinese designs and what they learn from working with their foreign partners. China has little experience of the boiling water reactor (BWR), but will be watching developments and eventual deployment of this technology in China cannot be ruled out.

There are other issues worth noting with regard to nuclear in China. On uranium supply, it is clear that the Chinese resource base is poor and that it will become a significant net importer. Production is believed to be no more than 750 tonnes of uranium (tU) per annum and with requirements approaching 2000 tonnes each year from current reactors, inventories are being drawn down and some initial contracts with overseas suppliers entered into. On spent fuel management, China has adopted a closed cycle, but doesn't yet have the facilities to reprocess the spent fuel and then utilise MOX or reprocessed uranium. Spent fuel is being moved from reactor sites and will eventually be reprocessed, a strategic decision that seems to be based largely on China's uranium weakness.

Concerning other fuel cycle facilities, China has growing enrichment capacity supplied by Russia, both gas diffusion and centrifuges, currently amounting to about 1.5 million separative work units (SWUs) per annum. There are also fuel fabrication facilities for both PWR and CANDU fuel, which can be expanded to meet local need. Therefore over a period of time, China aims to become increasingly self-sufficient in its fuel cycle needs, with the obvious exception of uranium, a notable weakness which is difficult (and certainly uneconomic) to plug.

Within reactor development, China has already exported one 300 MWe PWR to Pakistan, with another one under construction. Tsinghua University in Beijing has also developed a small high temperature gas cooled reactor (HTGR), which is currently operating and a full-scale prototype is under construction. Although it will be a long time before China will be a major nuclear exporter, it sees itself as a friend of other developing countries and will follow Korea on seeking nuclear cooperation with these. In particular, it has been acquiring uranium interests in developing countries and has signed a nuclear cooperation agreement with Australia to allow imports from there.

China will therefore become an increasingly important factor in the world nuclear industry, just as it is becoming in other key sectors of the global economy. The continuation of this is, however, dependent on China avoiding an economic meltdown, following the years of incredible growth. There are certainly signs within China of the existence of a bubble economy but most commentators are confident that the government can manage the eventual slowdown to a slower and sustainable economic growth rate. This should afford a good backdrop for steady nuclear growth, so that nuclear plays a more substantial part in national energy supply.

India: tomorrow's giant?

Foreign companies flock to China in the hope of securing orders in what may be the world's greatest nuclear power programme but prospects are admittedly rather thin elsewhere, certainly in the immediate term. The United States is a possibility but persuading tough-minded utilities to invest in something which requires a significant wad of cash over many years, but sells into uncertain power markets, is difficult. The fifth Finnish reactor is a shining beacon in Europe but political cowardice in facing the long-term energy supply situation means that there is still as much talk of shutting down existing reactors as building new ones. This ethos applies rather less to the new entrant countries to the European Union (EU), with relatively stronger political support for nuclear, while Russia also has ambitious plans for new nuclear plants. Yet there remains one other large country with a rapidly growing nuclear sector.

Among all the statements of 'keeping the nuclear option open' (in other words, we'll try to do without it, but if this fails we may have to reluctantly swallow our medicine) there is at least one country in the world with a clear, strong political commitment to nuclear power. The fact that it also has the second largest population (and may at some point soon have the largest) and is also is a vibrant democracy should cause more people to stand up and take notice.

At the Johannesburg Earth Summit in October 2002, India affirmed its commitment to nuclear power as an essential element of its sustainable development strategy. This was sufficient to see off anti-nuclear resolutions designed to constrain the industry everywhere. This political statement in favour of nuclear is, indeed, even stronger than that of China, where government support has wavered somewhat in the past and has only more recently apparently been won over.

From only 3.8 GWe of net nuclear generating capacity today, accounting for only 2.6% of electricity generation, India plans to increase this to 20 GWe in 2020. This will prove challenging, but is only a drop in the ocean compared with its longer-term vision. India recognises that its economic growth aspirations (a similar rate of growth to that achieved by China in the recent past) cannot be accomplished without exploitation of its significant coal resources (as per China). So up to 2020, new coal plants will continue to dominate the growth of electricity generating capacity. The environmental consequences of coal are, however, well-recognised so it is the intention beyond 2020 to increasingly rely upon nuclear for power generation, with capacity heading towards 100 GWe and then well beyond. The stated aim is to achieve 25% of electricity supply through nuclear by 2050.

There is one major constraint on India's plans, in that, as things stand, they need to be largely achieved by self-sufficiency in the nuclear fuel cycle. This is because of India's refusal to sign the 1970 Treaty on the Non-Proliferation of Nuclear Weapons (NPT), which it sees as discriminatory as it does not accept India as a nuclear weapons state. China exploded its first nuclear weapon in 1964 and India in 1974, but between these dates, the Treaty came into force. The NPT only allows India the option of renouncing nuclear weapons and taking what it regards as the inferior status as a non-weapons state. Owing to its security concerns about China and Pakistan, India refuses to do this and seeks equal recognition with the five weapons states, the United States, Russia, United Kingdom, France and China.

In an effort to induce expanded participation in the NPT in 1992, an informal club of nations called the Nuclear Suppliers Group (NSG) decided to prohibit all nuclear commerce with those which have not agreed to accept full-scope safeguards on their nuclear materials. This effectively requires countries to accede to the NPT if they are to participate in nuclear commerce. India's response has been to intensify its efforts to achieve self-reliance with a dual policy of maintaining a small nuclear deterrent while pursuing peaceful nuclear power on a major scale.

India's self-sufficiency extends from uranium exploration and mining through fuel fabrication, heavy water production, reactor design and construction, to reprocessing and spent fuel management. It has a small fast breeder reactor and a much larger one is now under construction. It is also developing technology to utilise its abundant resources of thorium as a nuclear fuel. Its uranium resource base is relatively weak and it is the denial of access to the international market that constitutes the major challenge to be overcome. It is in the area of fuel supply that the situation is most pressing, as India's current reactors have been short of fuel and unable to achieve high capacity factors.

Plans for the future are in three stages. Stage 1, already reached, employs pressurised heavy water reactors (PHWRs) fuelled by natural uranium to generate electricity and produce plutonium as a by-product. Stage 2 uses fast breeder reactors (FBRs) burning the plutonium to breed U-233 from thorium. Finally, stage 3 foresees advanced heavy water reactors (AHWRs) burning the U-233 with thorium, obtaining about 75% of their power from the thorium.

India's reactor plans have in the past been badly affected by many delays and poor operating load factors, which today mean that some outside observers doubt its ability to achieve the current plans.

However, it is clear that both the construction programme and reactor operations improved considerably in the 1990s – reactor capacity factors have reached 85% when fuel has been available. India's recent economic performance has also been much-improved, in contrast to the previous dismal record. The 17 current reactors of aggregate 3.8 GWe capacity include two small boiling water reactors (BWRs) from General Electric in the United States and two small Canadian PHWRs. (These four units predate the NPT, which put an end to trade with non-signatory countries.) The remainder are 11 local 200 MWe PHWRs, based on a Canadian design, and two larger 490 MWe examples. The share of nuclear in Indian power generation should rise above 5% with the existing construction programme. There are now three further 200 MWe PHWRs included in this, plus two VVER-1000s from Russia. The export of Russian technology escapes the NSG controls because the project is long-standing, predating the establishment of NSG. It rather stands outside India's 3-stage plan for nuclear and is an attempt to expand nuclear generating capacity more rapidly. Construction of all these reactors is running on or ahead of schedule. In addition, work has started on the first 500 MWe prototype FBR, which is expected to enter operation in 2011.

To achieve the target of 20 GWe of nuclear capacity by 2020, a major construction programme will have to take place from 2010 onwards. This will comprise a mix of higher capacity (up to 680 MWe) PHWRs, 500 MWe FBRs, 1000 MWe VVER-1000s (up to six are envisaged at the Kudankulam site, plus initial AHWRs. It is generally believed that such a programme cannot be achieved so quickly without an end to India's isolation in nuclear fuel and technology – continuing with self-sufficiency will delay the programme. There will also have to be major investments made in the electricity grid across India, where it is currently difficult to absorb the power from such large generating units.

The Indian nuclear sector is well-organised under the direct control of the Department of Atomic Energy. The Uranium Corporation of India Ltd operates underground uranium mines and is developing the extensive thorium resources. These account for about one quarter of the world total and contrast with the poor position in uranium, with only 54,000 tonnes of reasonably assured resources. There are plans to increase uranium production from the current level of around 230 tonnes per year, but this will take time. The Nuclear Fuel Complex at Hyderabad is responsible for uranium refining and conversion and also for fuel fabrication for all reactors. The Nuclear Power Corporation of India Ltd builds, owns and operates the nuclear power plants. The research and development facilities are impressive, including the Bhabha Atomic Research Centre near Mumbai and the Indira Gandhi Centre for Atomic Research near Chennai. There are clearly many technical challenges to be overcome in moving towards a thorium-based fuel cycle, but this is a necessity given India's current isolation.

The question remains as to whether this isolation can be ended. Hopes that India (plus Pakistan and Israel) will eventually sign the NPT are entirely misplaced. An overhaul of the entire world non-proliferation regime is clearly now required, taking account of current realities. This is only at the very early stages but the recent agreement proposed with the United States provides at least some hope that India may gain access to overseas technology and materials well before 2020, allowing it to fine-tune its strategy. This agreement faces a number of obstacles, both domestically with the Communist Party (which has an effective veto over the policies of the United Progressive Alliance coalition), within some NSG members and even in the US Congress. But to India's credit, it has always been scrupulous in ensuring that its weapons material and technology are guarded against commercial or illicit export to other countries. This contrasts with the position of Pakistan, whose enrichment technology has been diverted to Iran, North Korea and Libya.

Strong political backing for nuclear power is clearly a necessary precondition for its success. India's commitment is praiseworthy, in particular that it is linked to a long-term environmental policy. The consequences of a 1 billion plus population reaching much higher levels of per capita power consumption via fossil fuels will be considerable and a shift towards nuclear beyond 2020 provides an answer to this. Most countries that envisage long-term energy scenarios with low carbon emissions incorporate rather unrealistic assumptions for the role of renewables in power generation, but the Indian plans are entirely feasible if the technical challenges can be overcome. If its current isolation from nuclear trade can be ended, confidence in its achievements will be further enhanced.

Argentina & Brazil: a Latin revival?

Nuclear power achieved early footholds in Latin America, in both Argentina and Brazil, but it can be argued that it then rapidly entered a blind alley. Both countries stopped their programmes after only two reactors, with the previous expansive plans put on hold or cancelled. Reactors remain uncompleted in both countries and fuel cycle developments have accordingly been constrained, despite well-educated technical staff and some promising national research developments. Both countries have suffered from both political and economic problems, which have clearly not helped, but there are now signs that these have been overcome. The economies of both Argentina and Brazil are both now growing rapidly and electricity demand is rising sharply. There are substantial fears about shortages of generating capacity, so is a rebirth of nuclear very likely, to accompany the favourable developments underway in Asia, the United States and some European countries?

Other important features of both Argentina and Brazil include their desire to have full fuel cycles supporting their reactor programmes and not be dependent on supplies of nuclear fuel from overseas. Argentina's uranium resources are comparatively poor, but Brazil's are extensive (in the world's top ten in magnitude), if rather low grade. Both have sought to develop these to at least satisfy their domestic reactor requirements and, in the case of Brazil, possibly to export too. Conversion, enrichment and fuel fabrication technology (and some facilities) have also been developed with enrichment (in the case of Brazil in early 2005) attracting non-proliferation concerns. Yet both Argentina and Brazil are parties to the Treaty on the Non-Proliferation of Nuclear Weapons (NPT) and there has been a Argentine-Brazilian Agency for the Accounting and Control of Nuclear Materials (ABACC), set up with full-scope safeguards under International Atomic Energy Agency (IAEA) auspices since 1994. Both are members of the Nuclear Suppliers Group (NSG) but have not yet signed the Additional Protocol in relation to their safeguards agreements with the IAEA.

Considering Argentina first, the two operating reactors Atucha 1 and Embalse satisfy about 9% of the country's electricity requirements, which are now growing rapidly following the recovery from the economic 'meltdown' in 2001-2. About one third of the requirements come from hydro and the remainder from fossil fuel generating modes. Argentina decided at an early stage to go for heavy water reactors fuelled by natural uranium and invited bids from Canada and Germany in the mid 1960s. A Siemens KWU bid was accepted and Atucha 1 entered operation in 1974, located near Buenos Aires. It has net capacity of 335 MWe and is unusual amongst heavy water reactors in having a pressure vessel and now uses slightly-enriched uranium fuel (0.85% U-235), which has doubled the burnup and cut operating costs. The fuel is obtained by blending imported enriched uranium with natural uranium.

A second feasibility study in the late 1960s resulted in the selection of a CANDU 6 reactor from AECL of Canada for a second site at Córdoba, about 500 km from Buenos Aires. This included a technology transfer agreement and entered operation in 1984. It has net capacity of 600 MWe and, like all other CANDUs (to date), runs on natural uranium fuel.

Beyond this, a government decision in 1979 planned four more units to come into operation in 1987-97. Construction on only one of these was started, namely Atucha 2, a larger (690 MWe) version of the first unit, under a joint venture between KWU and the Argentine Atomic Energy Commission (CNEA), which coordinates all nuclear activities there. Work unfortunately proceeded very slowly with this, largely owing to lack of funds, and was suspended in 1994 with the plant an estimated 80% complete. Construction has since recommenced, but both money and expertise (to complete a unique design) have been in short supply. Completion has always seemed 'five years away' and that remains the position today.

Beyond the constrained power reactor programme, Argentina's main success has been in exporting research reactors (there are five operating in the country itself, indicating the high education standards there), including the new Lucas Heights reactor in Australia. It has produced only a cumulative 2500 tU from open pit and heap leaching uranium but has a uranium conversion company (Dioxitek) and fuel fabricator (Conuar), the latter of which has been successful in achieving export orders in non-nuclear metal finishing. There is a heavy water plant, more than sufficient for domestic needs, and some experimentation with innovative gaseous diffusion enrichment technology.

Turning to Brazil, the position has many similarities. The economy is now growing well with electricity demand booming, while 90% comes from hydro resources, mainly located a long way from the major demand area in São Paulo and Rio de Janeiro. This has worrying security of supply implications and is being addressed by plans for more fossil-fuel generating plants. The two operating nuclear reactors, Angra 1 and Angra 2, provide only about 4% of the nation's requirements.

Westinghouse won the bid for the first reactor on a turnkey contract and Angra 1, a 626 MWe PWR, opened in 1982 after a 10-year construction period. In 1975, the government signed an agreement with the former West Germany for the supply of eight 1300 MWe PWRs over 15 years, with the first two to be built immediately with equipment from Siemens KWU and the remainder a 90% local content under a technology transfer agreement. Economic problems meant that construction of the first two units was severely interrupted, and Angra 2 didn't come into operation until 2000. It has operated very well since then (in contrast to Angra 1 which had only a 25% capacity factor in its first 15 years of operation) but its twin, Angra 3, has barely been started, even though 70% of the equipment has been paid for and is onsite. It apparently costs $20 million per annum to maintain this in tip-top shape for future use, but plans to complete the reactor have kept being shelved. It has now been announced that the reactor will be completed (it is estimated that it will take 5-7 years and cost around $2 billion) while a further four reactors are planned from 2015.

Fuel cycle activities in Brazil are now all under the state-owned holding company Indústrias Nucleares Brasileiras SA (INB). This has developed the Lagao Real heap-leaching uranium operation with capacity of 400 tU per annum and there are plans to expand operations there and elsewhere, in order to capitalise on Brazil's extensive reserves. Overseas partners (including the Chinese) have been sought, but even with the recent strong increase in world uranium prices, the economics may be marginal. Conversion and enrichment have been purchased from abroad but Brazil has developed its own centrifuge enrichment technology, apparently similar to Urenco's, initially for naval reactors. This has now been reoriented to supplying the Angra reactors and it is believed that capacity will eventually be around 200,000 SWU per annum, sufficient for the two operating units. There is also a fuel fabrication plant, designed by Siemens, adequate for domestic needs.

Looking to the longer-term future, Brazil is involved in a wide range of nuclear R&D in five nuclear research centres, and is a member of the Generation IV International Forum (GIF) and also involved in the IAEA International Project on Innovative Nuclear Reactors and Fuel Cycles (INPRO) programme.

What can we conclude about nuclear prospects in these important countries with long traditions in the industry? The key is clearly the completion of both Atucha 2 and Angra 3, which would kick-start the supply industries that have been starved of work in recent years. Firm commitments to complete the reactors by a set date in one country is likely to be influential in the other, as politicians and industrialists in both countries watch each other very carefully and there is a good tradition of cross-border cooperation in nuclear matters. As such, the situation is rather similar to that in the former Soviet Union, where the first task has always been to complete those reactors that were partially completed in 1990. This is gradually happening and both the Russian Federation and Ukraine are now looking forward to more expansive reactor-building plans, which contain a greater degree of realism than in the recent past.

One concern, however, is the decline in the number of experienced nuclear staff in both Argentina and Brazil. The reduction in nuclear education in the universities has, as elsewhere, led to a significant ageing in the nuclear workforce. Assistance will be required from abroad, but the position is not so good in North America and Europe either, especially if these regions start building nuclear plants once again in great numbers. Nevertheless, the first stage is to get firm government support to complete the reactors – given their electricity supply problems and need to curb carbon emissions, this must remain the industry's first target.

Development of the fuel cycle activities also depends on the reactor completions, as they are solely (at present) dedicated to satisfying domestic requirements. This is even if the costs will be higher than importing from abroad – security of supply is important in both countries as they have relatively weak domestic energy resources. It is unlikely that Brazil will become a major exporter of enrichment and it may face pressure to close the facilities for non-proliferation reasons, if the current moves sponsored by IAEA towards regional enrichment and fuel reprocessing facilities bear fruit.

What is really needed, however, is a strong nuclear revival in Europe and North America. While these leading regions are not building reactors, there is no strong feeling in Latin America that nuclear is the way to go and it may be difficult to stimulate the completion of Atucha 2 and Angra 3. The longer they remain uncompleted, the more difficult it will be to find the right human resources.

Elsewhere there has been some talk of a reactor programme in Chile, the most successful country in Latin America in an economic sense, but this again needs a kick-start from the existing major nuclear countries.

Japan: where to now?

Nuclear power has been a cornerstone of Japanese energy policy since the mid 1970s and now accounts for about 30% of electricity production. This is despite being the only country to have suffered the devastating effects of nuclear weapons in wartime. Japan lacks significant domestic energy resources, having to import some 80% of its requirements. It is therefore very vulnerable to disruption in international fossil fuel markets, so has seen nuclear as a counterweight to this. The geographical and commodity vulnerability became very clear during the oil shocks of the 1970s and is reinforced by the more recent price escalations. Uranium has to be imported but it accounts for a relatively minor cost of power production at nuclear plants and can be easily stockpiled. Japanese utilities tend to hold 4-5 years of their annual uranium requirements in inventory and the 55 nuclear reactors now in operation have clearly greatly enhanced Japan's energy security.

More recently, however, the nuclear programme has slowed somewhat, with new reactors either being delayed or cancelled. There are several reasons for this. One is that forecasts of electricity demand growth in Japan have been regularly reduced, as the national economy has slowed. This has affected the plans of the ten major electricity utilities to introduce new generating capacity of all sorts, not just nuclear.

Secondly, public support for nuclear power in Japan has been eroded in the last few years due to a series of accidents and scandals. The accidents were the sodium leak at the Monju fast breeder reactor (FBR), a fire at a waste bituminisation facility connected with a reprocessing facility, and the criticality accident at a small fuel fabrication plant, which claimed two lives. None of these accidents were, however, in mainstream civil nuclear fuel cycle facilities. The steam accident at Mihama costing five lives (although at a non-nuclear part of the plant) and the earthquake at Kashiwazaki-Kariwa (although no injuries resulted, despite a seismic rating below the actual quake) have caused further concern. The scandals included the falsification of quality control data on a shipment from Europe of mixed oxide (MOX) fuel for light water reactors and the poor documentation of equipment inspections at Tokyo Electric Power Company's (TEPCO's) reactors, which extended to other plants. While the issues were not safety-related, the industry's reputation was sullied.

The industry has worked very hard on public acceptance for many years but the reputation for good safety and openness in dealing with the public has taken a knock. This has made it more difficult to get approval for new reactor sites and for local people to accept developments such as the introduction of MOX fuel in existing reactors. The earthquake affecting the Kashiwazaki-Kariwa plant in July 2007 has also had an impact on public confidence. Although the plant structure withstood the impact of an earthquake more severe than it was originally designed for, the seven reactors on the site are likely to remain closed for a lengthy period.

Finally, the question of what to do with nuclear waste has remained the subject of lively debate. Japan has embraced the policy of reprocessing spent nuclear fuel, initially at facilities in Europe but eventually at its own reprocessing plant, now nearing completion. The costs of this and also the wisdom of recycling the separated uranium and plutonium in current reactors have been questioned, which has created an additional climate of uncertainty surrounding the industry.

Japan imported its first commercial nuclear power reactor, Tokai 1, from UK. This was a 160 MWe gas-cooled reactor, which operated

1966-1998. After this unit was completed, only light water reactors (LWRs) – either boiling water reactors (BWRs) or pressurised water reactors (PWRs) – have been constructed. In 1970, the first three such reactors were completed and began commercial operation. There followed a period in which Japanese utilities purchased designs from US vendors and built them with the cooperation of Japanese companies, who would then receive a licence to build similar plants in Japan. Companies such as Hitachi Ltd, Toshiba Corporation and Mitsubishi Heavy Industries Ltd developed the capacity to design and construct LWRs by themselves. By the end of the 1970s the Japanese industry had largely established its own domestic nuclear power production capacity and today it exports to other East Asian countries and is involved in the development of new reactor designs likely to be used around the world. For example, Mitsubishi has submitted an application for standard design certification for its US-APWR to the US regulatory authority.

Due to reliability problems with the earliest reactors, they required long maintenance outages, with the average capacity factor averaging 46% over 1975-77. (By 2001, the average capacity factor had reached 79%.) After relatively poor operating performances of the initial reactors, the *LWR Improvement & Standardisation Programme* was launched by the Ministry of International Trade and Industry (MITI) and the nuclear power industry. This aimed, by 1985, to standardise LWR designs in three phases. In phases 1 and 2, the existing BWR and PWR designs were to be modified to improve their operation and maintenance. The third phase of the programme involved increasing the reactor size to 1300-1400 MWe and making fundamental changes to the designs. These were to be the Advanced BWR (ABWR) and the Advanced PWR (APWR).

Japan has progressively developed a complete domestic nuclear fuel cycle industry, based on imported uranium. There is a small uranium refining and conversion plant and while most enrichment services are still imported, Japan Nuclear Fuel Ltd (JNFL) operates a commercial enrichment plant with eventual capacity planned to be 1.5 million SWU per annum. There are several fuel fabrication facilities, and permission has recently been granted for a MOX fuel fabrication facility. This decision is seen as a significant step to closing the fuel cycle in Japan, as the programme has previously relied on European reprocessing and fabrication facilities. Operation is expected in 2012. At Rokkasho-mura, there is a major complex including the enrichment facility, low-level waste (LLW) and high-level waste (HLW) storage centres and a 800 tonnes per annum reprocessing plant. The planned MOX plant will also be located here. Originally the concept was to use the separated plutonium from reprocessing in FBRs, making Japan virtually independent regarding nuclear fuel, but this proved uneconomic in an era of abundant low-cost uranium. Development therefore slowed and the MOX programme shifted to thermal LWR reactors.

There are strong reasons to believe that nuclear will remain in its central position in Japanese energy policy for many years to come. Indeed, it is conceivable that it may become even more important. In July 2005 the Japan Atomic Energy Commission reaffirmed policy directions for nuclear power in Japan, while confirming that the immediate focus would be on LWRs. The main elements are that a "30-40% share or more" shall be the target for nuclear power in total generation after 2030, including replacement of current plants with advanced light water reactors. FBRs will be introduced commercially, but not until about 2050. Used fuel will be reprocessed domestically to recover fissile material for use in MOX fuel. Disposal of high-level wastes will be addressed after 2010.

In addition, Japan is heavily committed to achieving its Kyoto targets on greenhouse gas emissions and a strong role for nuclear is seen as central to this. It is already falling behind in its achievement of its targets for the protocol's initial 2008-2012 commitment period and appears to have few policies which will pull it back on track. Renewable sources now account for 12% of Japan's electricity generating capacity but the subsidies required to increase this are likely to have an adverse impact on electricity prices, always a 'hot' issue in Japan. The record on energy-saving and general conservation of resources is quite good, but the country remains very much hooked on substantial imports of fossil fuels, in line with other highly developed economies. This means that carbon emissions are continuing to rise. Finally, after a decade of little economic growth, it now appears that the Japanese economy is back on track and that electricity demand growth will be difficult to restrain. There has therefore been some talk of potential power shortages in the future unless substantial investments are made in the near future in new generating capacity, including nuclear.

Recent experience of building nuclear power plants has been very positive as construction periods have been reduced to four years and the latest evolutionary reactors, such as the advanced boiling water reactor (ABWR), have operated very economically. The recent increases in fossil fuel prices have also provided a significant spur to building new nuclear plants on economic grounds. The prime alternative to nuclear for baseload generating capacity is combined cycle gas turbine (CCGT) plants but economic viability crucially depends on the provision of a cheap gas supply, either by pipeline or (increasingly) through liquefied natural gas (LNG). Gas prices are much more localised that oil prices but the trend has been strongly upwards in recent times. Dependence on gas supply from potentially unstable countries is also a significant issue in Japan, just as in Europe.

Longer term, Japan is heavily involved in international research programmes to design the next generation of nuclear reactors, the so-called Generation IV. These will offer substantial advantages in safety, operating economics, proliferation-resistance and volumes of waste compared with the latest designs available today.

Given its continued concern about both energy security and greenhouse gas emissions, combined with a requirement to find fully economic solutions to energy requirements, Japan is likely to remain a major player in the world nuclear industry. In addition to building more nuclear plants, it is now a technology leader and fully capable of achieving export orders for equipment in other countries which are embracing nuclear, in particular China.

South Korea: a success story? With few indigenous energy resources, South Korean energy policy has always been driven by considerations of energy security and the need to minimise dependence on imports. Nuclear power has become an important element in this, from its beginnings in the late 1970s. Today, nuclear is big business in Korea, with 20 reactors in operation and significant plans for the future, a well-developed nuclear research establishment and companies now ready to supply the wider world industry. Yet this hasn't happened by accident, as it's the result of well-laid plans and good execution over an extended period of time.

The 'economic miracle' has taken Korea from developing country to advanced industrial nation very quickly. This performance is sometimes compared with Japan, but Korea was starting from a much lower base, without any significant industrialisation prior to the 1970s. Thus in some ways it is even more remarkable. Over the last three decades, the country has enjoyed 8.6% average annual growth in GDP, which has caused corresponding growth in electricity demand – up by a factor of ten from 36 TWh in 1978 to 365 TWh in 2005. The growth rate was maintained at more than 9% per annum from 1990 to the early years of the new century but is now expected to slow appreciably to 3-4% per annum by 2010. Per capita consumption in 2006 was 6400 kWh, up from less than 1000 kWh in 1980.

Nuclear generating capacity is now 17.7 GWe, representing nearly 30% of the total – the remainder being mainly coal and liquefied natural gas (LNG). But with high plant utilisation, it satisfies over 40% of annual demand. The first nuclear reactor to achieve criticality in South Korea was a small research unit in 1962 and ten years later, construction began of the first nuclear power plant, Kori 1, which reached commercial operation in 1978. After this there was a burst of activity, with eight reactors under construction in the early 1980s. The first three commercial units – Kori 1 & 2 and Wolsong 1, were bought as turnkey projects. The next six, Kori 3 & 4, Yonggwang 1 & 2, Ulchin 1 & 2, comprised the country's second generation of plants and involved local contractors and manufacturers. At that stage the country had six PWR units derived from Combustion Engineering (CE) in the United States, two from Framatome in Europe and one from AECL in Canada, of radically different design.

In the mid 1980s the Korean nuclear industry embarked upon a plan to standardise the design of nuclear plants and to achieve much greater self-sufficiency in building them. In 1987 the industry entered a ten-year technology transfer programme with Combustion Engineering (later acquired by Westinghouse) to achieve technical self-reliance, and this was extended in 1997. A sidetrack from this was the ordering of three more CANDU 6 PHWR units from AECL in Canada, to complete the Wolsong power plant. These units were built with substantial local input and were commissioned from 1997 to 1999.

In 1987 the industry selected the CE System 80 steam supply system as the basis of standardisation. Yonggwang units 3 & 4 were the first to use this, with great success. A further step in standardisation was the Korean Standard Nuclear Plant (KSNP), which brought in some further CE System 80 features and incorporated many of the US advanced light water reactor (ALWR) design requirements. It is the type used for all subsequent 1000 MWe units as well as the two cancelled partially constructed units in North Korea.

In the late 1990s, to meet evolving requirements, a programme to produce an Improved KSNP, or KSNP+, was started. This involved design improvement of many components, improved safety and economic competitiveness, and optimising plant layout with streamlining of construction to reduce capital cost. Shin-Kori 1&2 will represent the first units of the KSNP+ programme, and are expected to be among the safest, most economical and advanced nuclear power plants in the world. Beyond this, the Advanced Power Reactor 1400 (APR-1400) draws on CE System 80+ innovations, which are still evolutionary rather than radical. It offers enhanced safety and a 60-year design life, with initial construction cost expected to be about 10% less than the KSNP. This is important to maintain nuclear's cost competitiveness against coal. Construction is starting on the first APR-1400 units – Shin Kori 3 & 4 – and operation is expected by 2013.

The current 20 reactors in operation in South Korea are performing at very high levels and the safety record is excellent. In 2005 the capacity factor averaged 96.5%, the highest of any country. Licence renewals are currently being negotiated, to extend operating lifetimes by ten years, starting with Kori 1 and Wolsong 1. Power uprates of up to 5% are also envisaged for Kori 3 & 4, and Yonggwang 1 & 2.

Until April 2001 South Korea's sole electric power utility was Korea Electric Power Corporation (KEPCO). The power generation part of KEPCO was then split into six entities and all the nuclear generation capacity, with a small amount of hydro, became part of the largest of these, Korea Hydro & Nuclear Power Co Ltd (KHNP). KEPCO remains a transmission and distribution monopoly in public ownership but it is expected that the power generation companies, with the probable exception of KHNP, will eventually be privatised.

Government policy is to continue to have nuclear power as a major, indeed slightly increasing, element of electricity production. The Ministry of Science and Technology's (MOST's) plans project that South Korea should develop its nuclear industry into one of the top five in the world, with about 60% of electricity coming from nuclear by 2035. Another eight reactors are expected to come into operation by 2017, with four being the larger APR-1400s.

Within fuel cycle services, Korea has developed its own fabrication capacity to service its reactors. Korea Nuclear Fuel Company Ltd (KNFC) has capacity of 550 tonnes per annum for PWR fuel and 700 tonnes per annum for CANDU PHWR fuel. Uranium comes from Canada, Australia, and elsewhere – there will be a joint venture company with Kazakhstan to exploit the extensive uranium reserves there. Around 3500 tU are required to fuel the reactors in 2006, with enrichment requirements about 1.8 million SWU.

KHNP is responsible for managing all of its radioactive wastes. Fees are levied on power generation, collected by MOST and paid into a national Nuclear Waste Management Fund. Used fuel is stored on the reactor sites pending construction of a planned centralised interim storage facility by 2016, eventually with 20,000 tonne capacity. Long-term, deep geological disposal is envisaged.

The development of the Korean nuclear sector has now reached the point where it can be regarded as potentially a major exporter. Having developed its own standardised reactor designs based on imported technology, it is now in a position to export these to other countries. The Chinese market is clearly attractive and geographically close but it is more likely that Korea Power Engineering Company (KOPEC), the reactor design and engineering company, will concentrate on possible sales in new nuclear countries without strong existing ties, such as Indonesia, Vietnam and in South America, To this end, the Korean government has been signing nuclear cooperation agreements with such countries. The export effort is enhanced by having Doosan as a local, now privately-owned, contracting company, which is already achieving substantial orders on a worldwide basis for replacement major plant components such as steam generators.

One disappointment, however, has been the the Korean Peninsula Energy Development Organization (KEDO) project in North Korea. This was to build two 1000 MWe reactors, but construction was suspended late in 2003 and the project terminated in mid 2006 owing to the acute political problems with North Korea. Most of the fabrication of steam generators, pressure vessels and other equipment for both reactors has been completed and this equipment can now be sold off to other nuclear projects.

In conclusion, the progress of nuclear in South Korea stands out as a remarkable achievement. Sound energy planning has been an important element, with security of energy supply the prime initial motivation for 'going nuclear', as it was in France and Japan. The environmental advantages of a high nuclear share have now become increasingly important too, while the excellent operating performance of the reactors has ensured very sound economics at a time when LNG prices have been very volatile. Korea is unusual in having detailed plans for electricity generation needs running many years into the future. These will have to be adapted as demand growth varies, but it seems certain that nuclear will remain a major and probably increasing component. Learning from imported technology, the ability to design and build a distinctive class of reactor should mean that exports are a reasonable prospect over the next ten years. In this, Korea is perhaps 10-15 years ahead of China, which is still evaluating the merits of competing Western reactor styles.

Germany: is it really facing a phaseout?

Germany is clearly one of the leading nations in nuclear power, with a long record of both successful research and development work and also active development of operating reactors. Nevertheless, a phaseout of nuclear power is now enshrined in law, with strict limitations imposed on the operating lives of the reactors such that the last will close in the early 2020s. As such, this is much more serious as a phaseout plan than, for example, in Sweden, where the commitment is more long term and vague (yet has resulted in the two Barsebäck reactors closing). How has the threat to nuclear in Germany come about and what is likely to happen?

Germany's 17 operating nuclear power reactors, comprising 20 GWe of installed capacity, supply almost one third of the electricity. Many of the units are large and the last came into commercial operation in 1989. The operating performance has always been excellent, with many of the reactors achieving capacity factors above 90%. Six units are boiling water reactors (BWRs), 11 are pressurised water reactors (PWRs), while all were built by Siemens KWU. Thirteen of the reactors are licensed to use mixed oxide (MOX) fuel, using plutonium recycled from spent fuel, under reprocessing contracts with Areva in France and BNFL in the UK. When East and West Germany were reunited in 1990, all the Soviet-designed reactors in the east were shut down for safety reasons and are being decommissioned. These comprised four operating VVER-440s, a fifth one under construction, and a small older VVER reactor.

The other important elements to note in Germany's electricity supply are coal and wind power. Coal provides just over half of the country's electricity but some €2.5 billion per year is spent subsidising local coal mines. Yet arising from the Kyoto accord, Germany is committed to a 21% reduction of greenhouse gas emissions by 2010. Much of this has been achieved by shutting down energy inefficient old industries in the former East Germany, but there has been a major push towards wind power, backed by generous subsidies. Germany now has about half of Europe's installed wind generating capacity, amounting to about 25% of its total capacity but still only 5% of electricity supply.

The roots of the nuclear phaseout law in Germany run very deep, as there has long been a strong anti-nuclear movement in the country. The practical impact of this has been substantial. The Mülheim-Kärlich PWR (the only Babcock & Wilcox PWR ever built outside the United States) was the subject of licensing disputes since the 1970s and was stripped of its licence by judges in 1988, shortly after it began trial operation. It never operated thereafter. Similarly, a MOX fabrication plant was built at Hanau, but was never allowed to operate, so all MOX fuel has been imported. Finally, the waste disposal sites at Ahaus, Konrad and Gorleben have been subject to significant political delays, with transport operations attracting significant anti-nuclear demonstrations, well-covered by the media.

In 1998 a coalition government was formed between the Social Democratic Party (SPD) and the Green Party, the latter having polled only 6.7% of the vote. As a result, these two parties agreed to change the law to establish the eventual phasing out of nuclear power. Long drawn-out 'consensus talks' with the electric utilities were intended to establish a timetable for phaseout, with the Greens threatening unilateral curtailment of licences without compensation if agreement was not reached. All operating nuclear plants have effectively unlimited licences, subject to safety compliance, with strong legal guarantees.

In 2000 a compromise was announced which, while limiting plant lifetime to some degree, averted the risk of any enforced plant closures during the term of that government. In particular, the agreement put a cap of 2623 TWh on lifetime production by all 19 then operating reactors, equivalent to an average lifetime of 32 years (less than the 35 years sought by industry). Two key elements were a government commitment to respect the rights of utilities to operate existing plants, and a guarantee that this operation and related waste disposal would be protected from any "politically-motivated interference".

In June 2001 the leaders of the 'Red-Green' coalition government and the four main energy companies signed an agreement to give effect to this compromise. The companies' undertaking to limit the operational lives of the reactors to an average of 32 years meant that two of the smaller and least economic ones – Stade and Obrigheim – were shut down in 2003 and 2005, respectively, and that Mülheim-Kärlich would be decommissioned from 2003. It also prohibited the construction of new nuclear power plants for the time being and introduced the principle of onsite storage for used fuel.

The agreement was effectively a pragmatic compromise from both sides, which provided a basis and time for formulation of a sound national energy policy during a period when environmental and security of supply issues were becoming paramount. From the standpoint of the nuclear sector, the key benefit it gave was time – to continue operating the reactors in the hope that the 2002 revision to the Atomic Energy Act could be overturned by political change. Opposition leaders in the Christian Democratic Union and Christian Social Union (CDU/CSU) parties promised to reverse the decision as soon as they could but were thwarted in the 2005 elections. They did not achieve enough votes to avoid a coalition government and formed one with the still anti-nuclear SPD. Although the Chancellor, Angela Merkel, and other government leaders support the industry's desire for an extension of plant operations to 40 years at the very least, the nature of coalition politics has made this impossible. So German law is still for a phaseout as soon as the reactor generation allowances are used up.

The practical impact of this is now becoming serious, as four larger and highly economic reactors – Biblis A and B, Brunsbüttel and Neckarwestheim 1 – are expected to use up their remaining generation allowances soon after the next anticipated federal elections in autumn 2009. The utilities concerned are now using a variety of tactics to prevent such early closures, which are leading to legal actions between them and the government. These relate to the transfer of generation allowances from either shutdown reactors or those not scheduled to close until nearer 2020, to those under threat much sooner. Another possible tactic is to extend outages of these reactors so their allowances are not used up until after the next elections, in the hope that they will be saved by political change. Finally, if the worst comes to the worst and reactors are shut down, this apparently does not invalidate the operating licences, which remain effectively open-ended and only subject to regulators confirming that they meet safety standards. This opens the possibility of mothballing the reactors for subsequent restart should the political climate change, rather than decommission them (which in itself requires the issuing of a decommissioning licence by the regulator).

Nevertheless, these tactics avoid the central issue that the phaseout is enshrined national law and the German political system seems to lead itself to coalitions with insufficient power to change things. Reallocating generation allowances is rather like rearranging the deckchairs on the *Titanic* and a satisfactory solution will only be found for the long term if nuclear achieves incorporation in long-run German energy plans with popular support. This is going to be a significant battle – recent opinion survey results suggest that half the German population remains fundamentally critical of nuclear power and that, even in the overtly pro-nuclear CDU/CSU coalition, about one third of its supporters are against nuclear.

Probably the main hope is the anticipation that once the Germans are fully aware of the realities of a nuclear phaseout, they will react strongly against it. Renewable sources cannot possibly replace the power lost by closing the large nuclear plants, so more fossil powered generation capacity will be needed or electricity will have to be imported, conceivably from nuclear generation in France. Indeed, a January 2007 report by Deutsche Bank warned that Germany will miss its carbon dioxide emission targets by a wide margin, face higher electricity prices, suffer more blackouts and dramatically increase its dependence on gas imports from Russia as a result of its nuclear phaseout policy, if it is followed through. The bank estimated that 42 GWe of new generating capacity will need to be constructed by 2022 if the shutdowns proceed.

These realities seem rather obvious, but haven't yet been appreciated by enough German voters. As elsewhere around the world, there has been a general complacency about energy for many years, which is only now being replaced by detailed national policy debates. Where nuclear power is well-established, such as in the United States and United Kingdom, nuclear will play an important part in the debate, now not just relating to the continued operation of existing reactors but on new build too. Germany is a central element in the European Union (EU), where there is an important debate on energy concerning economics, environment and security of supply.

The impact on nuclear power as a whole if a German nuclear phaseout goes ahead would undoubtedly be very significant. With a clear leader in nuclear technology rejecting it, the chances of any new countries embracing the nuclear option look much slimmer. A country universally respected for its achievements in high technology engineering, shutting down large and highly economic generating units, is not the best advertisement for countries such as Indonesia and Vietnam, currently considering embarking on nuclear plans. It also seems highly unlikely that there would be substantial nuclear new build in the United States or elsewhere in Europe if Germany is phasing out reactors not substantially different from the new ones envisaged. Nuclear policy may still be nationally based but what is happening in other major countries is clearly still of some significance. The battle for continued operation of the German reactors is therefore a very important one.

Eastern Europe: a lively bright spot?

The outlook for nuclear power within the European Union (EU) has undergone an important change with the accession of new countries with a strong interest in nuclear. Of the ten countries which joined in 2004, five (Czech Republic, Hungary, Lithuania, Slovak Republic, and Slovenia) have operating reactors. Both Bulgaria and Romania, which joined in January 2007, also have nuclear reactors. In some cases, shutting existing older Soviet-era reactors was a condition of EU entry, but these countries retain a strong interest in nuclear and several have firm plans for new reactors in the future.

To some extent, the influence of these new members within the EU is beginning to act as a useful counterbalance to anti-nuclear forces in the 'old' members in Western Europe, particularly centred in Austria, Denmark and the Irish Republic. Indeed, the outlook for nuclear in Eastern Europe looks much better than further west, where Belgium, Germany and Sweden have phaseout policies and new build possibilities currently seem limited to Finland, France and perhaps also the United Kingdom.

In Bulgaria, six VVER reactors were built at the Kozloduy site, but the four VVER-440s have been closed as a condition of EU accession, leaving only the two larger VVER-1000 units in operation. A second site was originally chosen near Belene, also near the Danube border with Romania, with a view to building up to six large units. Construction of the first VVER-1000 unit started there in 1987, but was aborted in 1991 due to lack of funds. The government has now revived the Belene project and early in 2005, the government approved the construction of Belene as a 2000 MWe plant. Two consortia submitted bids to build the plant and in October 2006, the local utility NEK chose Atomstroyexport (ASE) over a Skoda-led consortium. Russia's ASE leads a consortium including Areva NP and Bulgarian enterprises in the project. The new units will be similar to those being built by ASE in China and India and the first is expected to come online in 2013. NEK will carry 51% of the project and is seeking partners such as Enel and CEZ, from Italy and the Czech Republic, respectively.

In the Czech Republic, construction of the Dukovany plant commenced in 1978, with four VVER-440 type V-213 reactors designed by Russian organisations and built by Skoda. These came into commercial operation 1985-87. In 1982 work started on the Temelin plant, with two VVER-1000 reactors. Construction was delayed and when it resumed in the mid 1990s, Westinghouse instrument and control systems were incorporated. The reactors started up in 2000 and 2002, with the upgrading having been financed by CEZ, partially from a loan from the World Bank. Both units have had significant ongoing technical problems with fuel and with turbines since they were commissioned. A further two units were originally envisaged on the site and the 2004 state energy policy envisages building two or more large reactors, probably at Temelin. Plans announced in June 2006 envisage one large unit at Temelin after 2020, with a second to follow soon thereafter.

In Romania, a five-unit nuclear power plant was planned at Cernavoda on the Danube River in the late 1970s. After considering carefully both Russian VVER-440 and Canadian CANDU technology it was decided to adopt the latter. Cernavoda was based on technology transfer from Canada (AECL), Italy and the USA, with CANDU 6 heavy water reactors. Construction of the first unit started in 1980, and of units 2-5 in 1982. In 1991 work on the latter four was suspended in order to focus on unit 1, responsibility for which was handed to an AECL-Ansaldo (Canadian-Italian) consortium. Unit 1 was connected to the grid in mid 1996 and entered commercial operation in December 1996. In 2000, the government decided that completion of Cernavoda 2 was a high priority and supplied some €60 million towards it. It was seen as the least-cost means of providing extra generating capacity for the country. Further finance was raised in 2002-03, with a €382.5 million package announced by the government, including €218 million from Canada. In addition, a €223.5 million Euratom loan was approved by the European Commission for completion of unit 2, including upgrades.

Unit 2, built by the AECL-Ansaldo-SNN (Societatea Nationala Nuclearelectrica) management team, came into commercial operation in October 2007. Efforts are also underway to resume work on unit 3, and SNN commissioned a feasibility study from Ansaldo, AECL and KHNP (South Korea) in 2003. It has now been decided to proceed with creating a project joint venture to complete both units 3 and 4 in a €2.2 billion project. A tentative schedule is for commissioning unit 3 in 2014 and unit 4 in 2015.

As a precondition for Slovak entry into the EU in 2004, the government committed to closing the two Bohunice V1 units due to perceived safety deficiencies in that early model reactor. The original date specified for closing them down was 2000, though subsequently 2006 and 2008 were agreed in relation to EU accession. The latter dates were set despite their recent major refurbishment, including replacement of the emergency core cooling systems and modernising the control systems. An upgrade programme on the two Bohunice V2 units is now underway to improve seismic resistance, cooling systems, and instrumentation and control (I&C) systems with a view to extending operational life to 40 years (2025). Areva NP is replacing the I&C systems progressively to 2008.

In 1981 construction of the four-unit Mochovce nuclear power plant was commenced by Skoda, using VVER-440/V-213 reactor units, but work on units 3 & 4 was halted in 1994. Units 1 & 2 have been significantly upgraded and the I&C systems replaced with assistance from Western companies. In October 2004 the government approved Italian Enel's bid to acquire 66% of the utility Slovenské elektrárne (SE) as part of its privatisation process. Enel's subsequent investment plan approved in 2005 involved a €1.88 billion investment to increase generating capacity and incorporating a plan to complete Mochovce units 3 & 4 – 942 MWe gross – by 2011-12. In January 2006 the government approved a new energy strategy incorporating these plans, with capacity uprates of 44 MWe gross at Mochovce 1 & 2 in 2007 and a further 18 MWe by 2012, and a 120 MWe gross uprate of both Bohunice V2 units by 2010. In February 2007 SE announced that it would proceed with Mochovce 3 & 4 construction later in the year, and that Enel had agreed to invest €1.8 billion on this with a view to operation in 2012-13. SE has already invested €576 million in the two units.

In Hungary, the four Paks VVER-440 units provide about 35% of the country's electricity but there are currently no additional reactors planned. In Slovenia, the only reactor is a 676 MWe Westinghouse-origin PWR, which provides over 40% of the small country's electricity. As in Hungary, there has been discussions about an additional nuclear unit, with recognition of the importance of nuclear power to both countries, but the planning horizon for these is likely to be beyond 2020.

A more interesting case is that of Poland, which is a large country where some 97% of the electricity is from burning coal. The Polish cabinet decided early in 2005 that for energy diversification and to reduce carbon dioxide (CO_2) and sulphur emissions, the country should move immediately to introduce nuclear power, with an initial plant operating soon after 2020. Poland had four 440 MWe Russian units under construction in the 1980s at Zarnowiec, but these were cancelled in 1990 and the components sold. In July 2006 Lithuania invited Poland to join with Estonia and Latvia in building a new large reactor in Lithuania, to replace the Ignalina units being shut down as a condition of EU entry. Polish participation would justify a larger and more economical unit such as a 1600 MWe European Pressurized Water Reactor (EPR). A 2006 feasibility study, undertaken on behalf of the three Baltic states, showed that a new plant costing €2.5 to 4.0 billion would be economically attractive and could be online in 2015.

The contrast between this wealth of activity in Eastern Europe and the lack of positive nuclear plans further west is indeed striking. It has several explanations. Certainly, these countries are already benefiting economically from EU membership and electricity demand is rising steadily. In itself, this brings with it consideration of energy policy and the available generation alternatives. Having just broken free of economic dependence on Russia, no-one wants to imperil energy independence by relying upon Russian oil and gas, nor import large quantities of power from neighbouring countries. The environment is also an important consideration, leading to coal being widely rejected as a serious option and the greenhouse gas avoidance merits of nuclear recognised. Renewable energy sources are advancing but cannot expect to provide the large quantities of power that these countries require, at least yet. The economic benefits of nuclear are also appreciated as the countries are proud of the operating performance of their older reactors. Indeed, they feel that there was never any real need for any of them to be shut down on safety grounds for EU accession, given the magnitude of safety upgrades undertaken. Finally, although public acceptance of nuclear varies by country, in general the overall climate appears to be much more favourable in Eastern Europe, with no substantial public opposition expected with the plans for new reactors. The approach to nuclear seems essentially pragmatic rather than ideological.

Russia: where's it heading?

The history of Russia's involvement in the nuclear has been told many times. In the 1950s and 1960s, it seemed to be taking impressive steps to contest world leadership in civil development of nuclear energy, but a technological arrogance developed, in the context of an impatient Soviet establishment. The Chernobyl accident tragically vindicated Western reactor design criteria, and the Soviet political structure, not up to the task of safely utilising such technology, eventually fell apart. There had to be significant changes and by the early 1990s, a number of Western assistance programmes were in place and helped to alter fundamentally the way things were done. Design and operating deficiencies were tackled, and a safety culture started to emerge.

Nevertheless, economic reforms following the collapse of the Soviet Union meant an acute shortage of funds for nuclear developments. Between the 1986 Chernobyl accident and the mid 1990s, only one nuclear power station was commissioned in Russia and since then, the reactor programme has continued to major on completing units under construction for many years. On the brighter side, by the late 1990s, exports of reactors to Iran, China and India were negotiated and signs were emerging that the domestic construction programme could soon be revived by better funding.

At the same time Russia became a significant exporter of nuclear fuel to the Western world. Much of this has been from inventories of materials built up in the past, particularly the downblended highly enriched uranium (HEU). Although Russia is not one of the largest uranium producers, it has formidable strength in centrifuge enrichment technology and plenty of capacity to serve export markets. Substantial export earnings have been achieved by exporting nuclear materials, which were particularly important in the early 1990s, when Russia had few alternative products to sell abroad.

Now things have changed and dramatically too. Major changes have been announced in the structure of the Russian nuclear sector together with ambitious plans for reactor construction at home, export plans to other markets and involvement in many international initiatives within the fuel cycle. At the same time, domestic fuel requirements and those for reactors built in other countries mean that Russia is no longer likely to play the same role of supplier to Western nuclear fuel markets, at least not on the same terms as before. Russia's future international role will essentially be to build on its reputation over the last decade as a reliable commercial provider of fuel-related services.

In April 2007, President Putin signed a decree to create a single vertically-integrated state holding company for Russia's nuclear power sector, separate from the military complex. The corporation, AtomEnergoProm (AEP), will include uranium production, engineering, design, reactor construction, power generation and research institutes in its several branches, so incorporating TVEL, Tenex, Rosenergoatom and Atomstroyexport. This is a sign of Russia's serious intent to be a major player in nuclear on a worldwide basis. The company has the clear potential, over the longer term, to be a player in the market not only of the magnitude of Areva, but combined also with Electricité de France (EDF).

On domestic power generation, it is important to note that Russian demand is rising strongly after more than a decade of stagnation and also that some 50 GWe of generating plant (more than a quarter of it) in the European part of Russia comes to the end of its design life by 2010. The most important point, however, is that it is now clearly Russian policy to export gas rather than use it for domestic power generation. Gazprom has cut back on gas supplies for electricity generation because it can supposedly make about five times as much money by exporting the gas to the West.

In September 2006, a target of nuclear providing 23% of electricity by 2020 was announced, thus commissioning two 1200 MWe plants per year from 2011 to 2014 and then three per year until 2020. This involves additional capacity of some 31 GWe by 2020 and providing some 44 GWe of nuclear capacity (net of closures) by then. A lower-growth scenario of adding only 2.4 GWe per year to 2020 was also mentioned, which would give around 37 GWe. The first stage must be to complete those reactors already under construction. A mid 2006 announcement pledged $665 million in 2007 towards completing Rostov/Volgodonsk 2, Kalinin 4 and Beloyarsk 4. Balakovo 5 & 6 seem to have been deferred while there is some uncertainty on Kursk 5, an RBMK design. It is apparently 70%

complete but still requires $750 million to finish. Beyond these units, initial plans for new reactors include three standard third-generation VVER reactors to be built: at Leningrad (two units as stage 2) and Novovoronezh (unit 6) to be commissioned 2012-13.

These plans are undoubtedly ambitious, particularly in the light of recent Russian performance at completing stated reactor plans. Delays have also been experienced in exported reactors, notably at Tianwan in China, with doubts expressed over project management skills. Nevertheless, the problem concerning finance for reactors in Russia, which has constrained plans for much of the period since 1990, would appear to be largely over. Difficulties in the future are likely to be similar to those discussed for elsewhere in the world, such as the availability of sufficient qualified staff and the capacity to manufacture the major reactor components in sufficient volume.

Two interesting developments are the plans for involvement with major aluminium smelters and the small floating nuclear power plants. In 2006 the major aluminium producer Sual (which has since become part of Rusal) signed an agreement with Rosatom to support investment in new nuclear capacity at Kola, to power expanded smelting there from 2013. Then in April 2007 Rosatom and Rusal, now the world's largest aluminium and alumina producer, said that they will undertake a feasibility study on an "energy metallurgical company comprising a nuclear power plant and an aluminium plant" in Russia's far east. Direct involvement with major power customers seems a logical way forward for new nuclear plants, on the lines of the fifth Finnish reactor, where the shareholders are big power consumers. Russia is also planning to construct seven further floating nuclear power plants in addition to one now under construction, on a barge to supply 70 MWe of power plus 586 GJ/h of heat to Severodvinsk, Archangelsk region. Five of the further plants will be used by Gazprom for offshore oil and gas field development and for operations on the Kola and Yamal peninsulas.

Russia still plans to close the fuel cycle as far as possible and utilise recycled uranium, and eventually also to use plutonium in MOX fuel. However, its achievements in doing this are so far rather limited. At present, the used fuel from RBMK reactors and from VVER-1000 reactors is stored (mostly at reactor sites) and not reprocessed. Used fuel from VVER-440 reactors, the BN-600 fast reactor and from naval reactors is reprocessed at the Mayak RT-1 plant at Ozersk (formerly Chelyabinsk-65) in the Urals. It started up in 1971 and employs the Purex process. The BN-800 fast reactor project, intended to replace the operating BN-600, may become international with Japanese and Chinese involvement. Construction has been continuously delayed by lack of funds, but it is now underway with hopes for a 2012 startup.

Russia's policy for building nuclear power plants in non-nuclear weapons states is to deliver on a turnkey basis, including supply of all fuel and repatriation of used fuel for the life of the plant. In October 2006, its bid for two VVER-1000 units for Belene was accepted by Bulgaria. ASE leads a consortium including Areva NP and Bulgarian enterprises in the €4 billion project. Since then, Rosatom has actively pursued cooperation deals in South Africa, Namibia, Chile and Morocco as well as signing a memorandum of understanding with Enel of Italy for cooperation on nuclear power projects in Eastern and Central Europe (where Enel has a major presence), using Russian technology. It is likely that Atomstroyexport will eventually build a second unit at Bushehr in Iran and two more in China. In early 2007, a memorandum of understanding was signed to build four more units at Kudankulam and other reactors elsewhere in India.

There has also been substantial activity in uranium production in recent times. In 2006, Russia produced some 3200 tU, but TVEL announced that this needs to increase to 7500 tU per annum by 2020 to match increased domestic demand. TVEL and Tenex have formed the Uranium Mining Company to consolidate their existing mining assets and to develop uranium deposits in Russia and in Kazakhstan, Ukraine, Uzbekistan and Mongolia. In October 2006 Japan's Mitsui & Co with Tenex agreed to undertake a feasibility study for a uranium mine in eastern Russia to supply Japan. This would represent the first foreign ownership of a Russian uranium mine. Finally, Tenex has also entered agreements to mine and explore for uranium in South Africa (with local companies) and Canada (with Cameco).

On enrichment, in addition to steadily rising capacity at the four Russian facilities, the International Uranium Enrichment Centre (IUEC) is being set up at Angarsk. This is part of President Putin's vision to provide new nuclear power states with assured supplies of low-enriched uranium for power reactors, giving them some equity in the project but without allowing them access to the enrichment technology. This fits in well with the International Atomic Energy Agency's (IAEA's) proposal for multilateral approaches to the nuclear fuel cycle and with the US Global Nuclear Energy Partnership (GNEP).

Finally on international cooperation, Russia has been the effective leader of the International Project on Innovative Nuclear Reactors and Fuel Cycles (INPRO) at the IAEA, which is working on next generation reactors. It has also recently signed a cooperation declaration with the OECD's Nuclear Energy Agency (NEA), bringing it much more into the mainstream of Western nuclear industry development. This agreement is expected to assist Russia's integration into the OECD.

There is therefore a lot more now happening in Russia. It seems that the shadow of Chernobyl and the difficult economic era immediately after the fall of the Soviet Union have finally been shaken off. The reorganisation of the industry along Western lines is encouraging and provides more credibility for the ambitious plans. Strong oil and gas export earnings appear to be giving the economic strength the Russian nuclear sector needs to fulfil the early promise it showed back in the early days of nuclear power.

New nuclear countries: where & when?

With the vastly increased attention being paid to nuclear power today, it is not surprising that a large number of new countries have been mentioned as possibilities for building and operating reactors. At present, there are 439 reactors operating in 30 countries around the world and Iran will add to the list once the Bushehr plant is completed. It is certain that nearly all the new reactors coming into operation in the period to 2020 will be located in these same countries, with only a few likely in new countries. Indeed, it is almost a precondition for extending nuclear power to additional countries that it is already thriving in those countries where it is already well-established. Nuclear programmes take time to get going and in the usual timescales with this technology, 2020 is not so far away. The period beyond 2020, however, opens up rather wider possibilities, and it is in the following period that we may realistically expect the number of nations operating nuclear plants to expand substantially.

The increased interest in nuclear in 'new' countries has been caused by similar factors to those important in those countries already with reactors. The improved operating and safety performance of the current reactors has transformed both perceptions about the economic performance of nuclear power and its level of popular public acceptance. The impact of the debate on greenhouse gas abatement and also the heightened concerns about energy security of supply has also been important.

Although it is easy for a country to announce that it is seriously considering a nuclear power programme, getting from A to B is rather easier said than done. Governments need to create the right environment for investment in nuclear power, including a professional regulatory regime and policies on nuclear waste management and decommissioning. This may prove challenging to many developing countries. There is also an obligation to satisfy international non-proliferation and insurance arrangements for nuclear power plants.

Of those countries that are serious candidates for achieving their first reactors, the most likely ones are in Europe and East Asia. Italy is today now the only G8 country without its own nuclear power industry, and is the world's largest net importer of electricity. Due to the high reliance on oil and gas, as well as imports, electricity prices are around 45% above the European Union (EU) average. Italy was, however, a pioneer of civil nuclear power and built several reactors which operated from 1963-90. Following a referendum in November 1987, provoked by the Chernobyl accident 18 months earlier, work on the nuclear programme was largely stopped. In 1988 the government resolved to halt all nuclear construction, shut the remaining reactors and decommission them from 1990. Italy was then largely inactive in nuclear energy for 15 years. In 2004, a new energy law opened up the possibility of joint ventures with foreign companies in nuclear power plants and also importing electricity from them. This resulted from a clear change in public opinion, especially among younger people favouring nuclear power for Italy. In 2005 Electricité de France (EDF) and the Italian utility Enel signed a cooperation agreement which gives Enel some 200 MWe from the Flamanville 3 EPR (1700 MWe) under construction in France, and potentially another 1000 MWe from the next five such units built there. In addition, Enel will also be involved in design, construction and operation of the plants. This will both enhance Italy's power security and economics, but also help in rebuilding Italy's nuclear skills and competence. Enel also has a 66% shareholding in Slovak utility Slovenské elektrárne (SE), which operates six nuclear power reactors in that country. These investments can be seen as an important precursor to Italy eventually building nuclear plants within its own territory.

Amongst the countries of the former Soviet Union, Belarus, Georgia and Kazakhstan have been mentioned as possibilities for nuclear reactors. Both Belarus and Georgia have poor power generation infrastructure and seek to reduce dependence on Russia. In Kazakhstan, the BN-350 fast reactor at Aktau on the shore of the Caspian Sea successfully produced up to 135 MWe of electricity and 80,000 m^3/day of potable water over some 27 years, until it was closed down in mid 1999. There are proposals for a new nuclear power plant near Lake Balkhash in the south of the country near Almaty.

Turkey has a long history of considering nuclear power but, as yet, no reactors have been built. In August 2006 the government said it planned to have three nuclear power plants with a total of 4500 MWe operating by 2012-15. This looks hopelessly over-optimistic but discussions are apparently underway with foreign vendors.

Turning to Asia, the most likely prospects are in Indonesia and Vietnam. Indonesia has been seriously considering nuclear build since the late 1980s and with a huge population of 242 million, is served by power generation capacity of only 21.4 GWe. The economy is growing strongly and a low reserve margin with poor power plant availability results in frequent blackouts. It is hoped that the first reactor will come online in the period 2016-18, while it is believed that Korean PWR technology is the favourite to be adopted. Vietnam has also been looking at the nuclear option for many years and in February 2006, the government announced that a 2000 MWe nuclear power plant would be online by 2020. A feasibility study for this is due to be completed in 2008 and formal approval will then be required to open a bidding process with a view to starting construction in 2011 and commissioning in 2017.

Elsewhere in Asia, Malaysia and Thailand are two countries with the level of economic institutional development to suggest that nuclear should be a viable generation option. Both have, in the past, shown little interest in nuclear plants, but a mixture of rapid economic growth and fears about over-dependence on natural gas in the generation mix is encouraging a new look. Energy plans are being revamped and nuclear may well become part of future strategies.

In South America, Chile is probably the most likely country to join Argentina and Brazil in building nuclear reactors. Economic performance has been the best in the region over the recent past but Chile is heavily dependent on energy imports. In February 2007, the Energy Ministry announced that it was beginning technical studies into the development of nuclear power.

There are a large number of countries in Africa and the Middle East now showing interest in nuclear power, but it is wise to be cautious in evaluating their plans. In some cases such as Egypt and Morocco, the desalination of seawater is an important objective as well as power generation. The International Atomic Energy Agency (IAEA) and Western vendors are working with various countries to assist them with their plans, but it is unlikely that more than a handful of new reactors will be seen in these areas before 2020.

The size and stability of local electricity grids is clearly a major issue in many developing countries, especially as major reactor designs are getting progressively bigger, to 1500 MWe and beyond. The pebble bed modular reactor (PBMR) being developed in South

Africa would be very suitable for much smaller grids, as it is planned at 170 MWe, but is unlikely to be generally available before 2015. Another issue is the ability of such countries to regulate nuclear power in an appropriate manner and to put in place sound waste management and decommissioning policies. Many of them already have research reactors operating for many years but power reactors substantially increase the magnitude of the issues to be addressed. Finally, there are concerns that some countries may be considering nuclear power to assist in a clandestine weapons plan, as Iran is being accused of doing. The Global Nuclear Energy Partnership (GNEP) proposed by the United States and complementary ideas from Russia and the IAEA, whereby countries can receive secure deliveries of nuclear fuel in return for abandoning any idea of developing sensitive enrichment or reprocessing facilities, hope to address such worries.

Ultimately it seems rather unlikely that many of these countries will obtain power reactors before the longer-term future of nuclear is decided in the more developed nations. For example, it seems highly unlikely that a country such as Malaysia will be building nuclear reactors at the same time as Germany is shutting down quite similar reactors, fully capable of operating economically for another 20 years. Similarly, unless the United States and some European nations, such as the United Kingdom, start building new reactors again, to prevent the nuclear share in electricity supply falling sharply, it is hard to see many developing countries doing so. So if the nuclear renaissance is to demonstrate real substance and not evaporate in a cloud of hot air, it is essential that new reactor orders be very soon achieved in the leading nuclear nations.

2. NUCLEAR NEW BUILD AND ECONOMICS

Introduction

It is clear that for the 'nuclear renaissance' to become a reality, there must soon be a significant number of orders for new nuclear plants throughout the world. As things stand, most of the growth prospects are in Asia and the former Soviet Union, but it is essential that the United States and some of the countries of Western Europe start building new reactors once again. Conditions are certainly far more favourable for this to happen than they have been at any time in the past 20 years, but everyone recognises that major nuclear projects must overcome a large number of hurdles which lie in their way.

Despite the increased attention now being paid to nuclear for environmental and security of energy supply reasons, it's clear that new nuclear plants must prove their economic credentials in the power markets of today. There may be some financial stimulus granted to nuclear because it emits few greenhouse gases and can reduce dependence on imported fossil fuels, but these cannot be counted upon. If it can be shown that a new nuclear plant will provide the cheapest baseload electricity over the long term, this is a very powerful argument in favour of selecting it. The relevant parameters in the equation may vary from country to country (and indeed by region within the bigger nations) and there may be no general case. But it is essential for the nuclear industry to be able to prove that it provides very cheap power, as well as environmentally benign and secure supplies. In many countries, governments seem set to allow new nuclear build and will support it with an improved licensing regime. But companies will be expected to bring forward projects without subsidies and in a way that provides good returns to their stakeholders.

This, in itself, represents rather a break with the past. Previous nuclear programmes were often creatures of the state, in an era when electricity supply was regarded as a public good, requiring state planning and control. While nuclear cannot break entirely free of governmental interest, competitive forces have been introduced to most electricity markets and nuclear has to respond to this. In particular, the past performance on reactor construction cannot be repeated. In many countries, notably the United Kingdom and United States, nuclear plants took far too long to build, weighed down by a mixture of technical, organisational and regulatory issues. In order to make any economic sense, the next set of reactors must ideally take no longer than four years to build, to allow electricity revenues to flow as quickly as possible. This timeframe has been achieved with the latest reactors in Asia, but now needs to be proved elsewhere.

That a country such as Finland can undertake a new reactor project, when nobody else in that region was doing so, has attracted a lot of attention. Perhaps they are crazy or possibly they have spotted something that nobody else has? In reality, the decision to go ahead with a new reactor there was a hard-nosed business decision, based on some special factors that have some relevance elsewhere.

There is a general expectation that much replacement or incremental generating capacity will be required in many countries in the period to 2030, so the underlying market for new nuclear build is good. The economic equation is now also much better understood, essentially boiling down to finding a way to cover nuclear's relatively high capital costs with low generating costs, once the plant starts up, as quickly as possible. There are now many studies which show that this is possible but there remain important issues, such as how to overcome the additional costs that are likely to accrue when the first units of new reactor designs are built.

The move to liberalisation of electricity markets throughout the world has posed certain challenges to nuclear – indeed, it was often asserted that this would kill off many of the existing nuclear plants. This hasn't, in fact, happened and it can be argued now that market liberalisation gave the shot in the arm that nuclear needed to improve its performance. Electricity liberalisation has taken on many different forms in different markets and is still a dynamic process, but new nuclear plants will have to fit in with its requirements. Financing has been brought up as a potential barrier to new nuclear build, on the basis that the projects are complex, long term and have experienced substantial delays in the past. It can be seen, however, that the main issue is structuring nuclear projects correctly, so that the risks and potential rewards are allocated correctly between the various stakeholders. Then the necessary financing should be forthcoming. Another issue is the ability of the supply side to provide all the plant materials and components if there is to be a substantial new nuclear building programme around the world. Given the lack of orders for new plants in recent years, it would be very surprising if significant spare capacity was currently available. Which indeed it isn't, but the trigger of a wave of new plant orders should allow a quick response.

Finally, to summarise, it can be seen that the challenges of new nuclear build can be overcome provided that governments have sufficient will to provide an appropriate policy and regulatory framework, and the right men and women come forward to lead the programmes from the industry side. Unless there is sufficient drive and enthusiasm in high places, the obstacles may prove too much to be faced and easier options selected.

New nuclear plants: can we ever again build in volume?

The construction of a new nuclear reactor in Finland is remarkable – particularly as it's a big reactor, rated at over 1600 MWe. Yet there is an acute paradox. Finland is an environmentally-conscious country, with a population of little over 5 million. To the chagrin of the anti-nuclear forces, it already has four operating nuclear reactors. On two sites too, one with Western BWRs and the other with smaller Russian VVERs. All have operated very well, with commendable safety records and high capacity factors which have given excellent economic returns to the owners. So why are many other countries not building new nuclear plants?

Let's apply the Finnish experience to the remainder of the European Union (EU). The population of the 27-state EU is in excess of 400 million, which is 80 times the Finnish level. So, multiplying, the EU should already be running 320 reactors and have the potential for 400 (80 x 4 = 320 and 80 x 5 = 400). Current reality, however, and potential numbers are below half of this (150 reactors are now operating in the EU-27). And even this is taking an optimistic view of a renewed reactor building programme in the EU before 2025, when some countries (Germany, Sweden and Belgium) seem politically set on phasing out nuclear power.

There are some unusual features about the Finnish experience. It is certainly rather cold in those northern latitudes and their industry is energy-intensive. On the other hand, they have good hydro resources and sit next to Russia, with its abundant oil and gas reserves. Coal can be imported easily on the international market. So there are plenty of alternatives for the Finns in power generation. They don't have to choose nuclear but have done so. Are they irrational when the rest of the world is sensible?

The answer is a strong no. The decision to go for a fifth Finnish reactor was a decision that was subjected to the utmost scrutiny, by government, investors and the general public. It contains an acute lesson for the world nuclear industry. This is that the main key to new nuclear build is economics. It was proved, beyond reasonable doubt, that a new nuclear reactor is the cheapest option for Finland. It can be argued that if you can make a strong economic case for new reactors in other countries, the battle will be won. Yet the industry has allowed itself to be diverted into trying to solve other perceived problems. So, a strong economic case for new build is still lacking today. We still cannot, in general, make it stand up to reasoned argument

The nuclear industry is fond of making excuses for its failures. In particular, it claims that everybody is against it. There is a list of issues which the general public attacks it for, then it makes often poor efforts to deal with these. There's plant safety, the fear of discharges to the environment, potential weapons proliferation, waste management issues (often said to be the biggest downside) and the decommissioning of plants (particularly the costs). Then there's nuclear material transport, the sustainability (or lack of it) of uranium resources and above all, the alleged regulatory burden.

There are clearly links from some of these to the economic dimension, particularly through regulation – some safety and discharge requirement levels cannot be justified on any reasonable risk-based cost benefit analysis. But this is just something the industry has to learn to live with. Other industries have their burdens too – in truth, with strong public scrutiny today, no industry gets an easy ride and the nuclear industry is too fond of hiding away thinking it is special, when it's not. The truth is that it's been forced onto the back foot too easily by opponents and wasted millions of man days in attempting to quell the opposition. In reality, all it has to do is to work harder on establishing an irrefutable economic case. Then the opposition will largely melt away.

So what is the economic problem and how have the Finns overcome it? From the point of view of currently operating nuclear plants, there is little concern as the existing stock of reactors could almost run forever. Extensions to operating lives are now the norm. Well-run nuclear plants throughout the world operate at high capacity factors and are also the safest. They have very low marginal costs (including operations, maintenance and fuel), usually guaranteeing excellent financial returns in liberalised power markets (but note the adverse UK example below). The marginal costs are also remarkably predictable and stable, which is a real bonus in this uncertain world. This also ignores the important energy security and environmental benefits of nuclear, as these don't yet have a clear economic dimension. So if you can get a good nuclear plant up-and-running, you could make a small fortune. But the problem is getting the plant going in the first place.

The potential economic problem for new nuclear plants is relatively simple to summarise. They have high construction costs but low operating costs. Combined cycle gas turbines (CCGTs) have the opposite cost profile (*i.e.* low construction costs and high operating costs) and have proved to be the generation option of choice in most electricity markets since the mid 1990s. The economic assessment involves balancing these two cost elements. The situation today is that nuclear's capital costs are believed to be, in most cases, still too high to meet the market requirements, which are very short term in nature in liberalised electricity markets. The construction cost handicap faced by a nuclear plant will, in most cases, be sufficient to destroy its chances. With relatively high commercial interest rates, the benefit of low fuel and operating costs in the distant future, compared with competing technologies, will be largely lost in financial calculations.

The extent of this handicap is, however, uncertain. It is generally believed that new CCGT plants can be brought on at $500-600 per kWe of installed capacity. Recent nuclear plants built in East Asia have had build costs in excess of $2000 per kWe. The fifth Finnish reactor, a European Pressurized Water Reactor (EPR) with advanced evolutionary features, was also quoted at similar levels – €3 million for 1600 MWe, which (with the weak US dollar) equates to well over $2000 per kWe. Major restart projects, such as at Browns Ferry in the United States and at the Pickering A plant in Canada, have costs quoted in the $1500 per kWe range. The nuclear reactor vendors claim that they can get below the $2000 per kWe level for new build, although with a fair wind behind them.

This is an important question – as we haven't been building new nuclear plants in the Western world for so many years, there is general uncertainty on the level of capital costs. Nevertheless, the key is not only capital costs. The Finns can go ahead at in excess of $2000 per kWe, but how?

The answer involves the structure of the electricity market and how the power is sold. British Energy got into financial difficulties in 2002 because it was a price-taker in a liberalised electricity market where even efficient, low marginal cost generators lost money. This was, however, an extreme case, where the level of prices and the acute short-term perspective meant that no rational company would invest in any new generating capacity, whatever the technology. That is unless there are government subsidies, as received by the renewables sector in the UK and elsewhere.

The Finnish situation is different as the plant investors will also be the main power customers and can therefore take a longer-term view. So the biggest challenge, other than minimising capital costs, is to find a way of allowing large capital investments to take place in liberalised power markets. Customers only want to pay the marginal cost for their electricity and not to finance new plants, either incremental or replacement capacity. Rather as they do with airline tickets, where competition sometimes drives prices down towards marginal operating costs, making investment in new aircraft unattractive.

Regarding lowering capital costs, there remains much to be done. Few countries could justify new build at the Finnish cost level. Pushing the aircraft industry example further, perhaps the industry needs to rationalise around a smaller number of vendors, each with a few standard designs, like Boeing and Airbus. These could be certified as safe in the United States, for example, then licensed quickly elsewhere. This would require a clean break with existing regulatory practice, but perhaps the industry needs to take the lead in pushing for this. We often say that nuclear is a mature industry, but perhaps it isn't – there may be a lot of life in the old dog yet.

New nuclear build: do the economics stack up?

The critics of nuclear power frequently assert that it is uneconomic. The industry evolved from post Second World War military nuclear programmes but the research and development work there did not always lead the subsequent civil sector down the optimum technical paths. Governments have more recently moved away from widespread intervention in energy markets. Electricity liberalisation comes in many guises, but the general trend is clear. The industry today recognises that all plants must demonstrate that they are cost-effective and that this must be achieved whilst still maintaining very high safety standards. Safety and the best economic operation tend, in any case, to go hand in hand.

Nuclear power plants constructed over the past 30 years have clearly demonstrated that they can meet the economic challenge posed by the market. Plant load factors have increased significantly, squeezing more output out of the same amount of capacity. Owners have found it worthwhile to invest in plant refurbishment and in capacity uprates, as marginal costs of generation from nuclear plants have been below those of most other generating modes. Marginal costs tend to be low, stable and predictable, in contrast to those of fossil fuel powered plants, where the volatile fuel prices are an essential part of the electricity cost. This has generated good profitability for nuclear plant owners, which has encouraged them also to seek operating licence extensions for many reactors. In the United States, an active market has developed in operating nuclear plants, as ownership is consolidated in a smaller number of companies, each with a significant commitment to its nuclear fleet.

World energy production and consumption have recently been growing at around 2% per annum and most projections see this continuing in the period to 2030. For example, the reference case in the 2006 edition of the International Energy Agency's (IEA's) *World Energy Outlook* projects that global primary energy demand will increase by over half in the period to 2030. This represents a growth rate of 1.7% per annum in the period 2000-2030.

It is also almost certain that the growth rate of electricity demand will exceed this, based on recent trends that favour the delivery of energy in this way. The proliferation of electric home appliances and strong underlying commercial and industrial demand growth underlie this assumption. In addition, it is estimated that there are currently still up to two billion people in developing countries who do not have regular access to electricity. Most (but not all of these) are living in rural areas and spreading electricity services to these 'unconnected' people is a prime objective of current development strategies. The IEA therefore projects that world electricity demand will almost double in the period to 2030, an annual growth rate of 2.5%. There is a marked contrast between the expected growth rates in different areas. In the OECD countries, growth of only just over 1% per annum is expected, whereas in both China and India, it should be approaching 5% per annum.

Given this background for world electricity demand, a significant amount of new investment in generating capacity is required. In the developed world, much of this will be replacement capacity but in the developing world, nearly all will be incremental. Much of this investment will be directed at satisfying local baseload requirements, even if one assumes an aggressive commercialisation of renewables and of distributed generation. Given this positive background, the nuclear industry sees a real opportunity for new nuclear build, on grounds of both plant economics and environmental sustainability.

As far as new electricity generating plants are concerned, the basic economic question can be presented quite simply. Are the lower and stable fuel costs of a nuclear plant compared with local competition from alternative generating modes sufficiently attractive to offset the higher initial capital costs?

It can now be assumed that investments in new nuclear generating capacity will only happen if the rate of return to investors is sufficiently high compared with other potential options to invest the funds and taking account of their appraisal of the risk level. The early stages of the shift to competitive electricity market regimes has attracted investors to favour investments where the need for capital is small and the construction times are short. Volatile fuel prices make such investments risky and where major power users seek long-term price stability, nuclear power plants offer a good solution. In addition, the uncertainties surrounding the character of power markets are not helpful for securing a rational and secure supply infrastructure built on large slugs of capital.

The key parameters in the economic competitiveness of new nuclear plants are their capital costs (strongly influenced by the construction time and the rate of interest to be paid on financing) and the fuel prices of gas and coal plants, the main alternatives in baseload generation. Various recent studies demonstrate that if new nuclear plants can be built with low capital costs and coal and gas prices are at current levels or higher, they can be cheaper on a levelised cost basis than the alternatives.

Now that the outlook for fossil prices is changing, the assumptions made in the reference cases in even very recent studies of generating plant competitiveness may now be regarded as too low, at least in the short term. Certainly the recent experience of oil, gas and coal prices suggests that they will remain volatile in the future, in contrast to key nuclear operating costs.

Everything possible needs to be done to bring the capital costs of new nuclear plants down to the level that the industry believes is possible, namely an overnight (before interest charges during the construction period) cost of below $2000 per kWe installed. It is clear, for example, that there are substantial economies of scale with nuclear plants, suggesting that reactor sizes should increase for economic reasons. There may be some constraints on this imposed by electricity grids, while another approach is to build several reactors on one site, which can bring major unit cost reductions. Standardisation of reactors and construction in series will also yield substantial savings. Learning-by-doing is regarded as potentially a significant way of reducing capital costs, both through replication at the factory for components and at the construction site. Simpler designs, possibly incorporating passive safety systems, can also yield savings as can improved construction methods. Initial plants of new designs, however, face substantial first-of-a-kind engineering (FOAKE) costs and may need some public assistance to become economic. Several schemes have been suggested for this.

The nuclear industry is working with regulators to develop licensing processes that have rigorous but predictable technical parameters and timescale, from design certification through to construction and operating licences. It is hoped that the industry can eventually move closer to the model set by the aircraft manufacturing industry and their corresponding regulatory authorities, where a limited number of designs once approved in one country are quickly licensed for use in all countries and can be operated by staff with internationally-recognised skills.

If fossil fuel use is significantly penalised by carbon taxes or emissions trading regimes, the competitiveness of new nuclear plants clearly improves. This is particularly so where the comparison is being made with coal-fired plants (because they are so carbon-intensive) but also applies, to a lesser extent, to gas-fired plants. The case for nuclear plants to be included in greenhouse gas avoidance schemes is very strong but the industry cannot necessarily rely upon much help from this to make its case in an economic sense. If it comes, it will be regarded as an added bonus. The industry should continue to strive to do everything possible to influence things over which it does have some direct influence, namely the costs of building and operating reactors.

The significant FOAKE costs attributable to new reactors must be recovered with early orders, but the lack of recent experience of building nuclear power plants in many countries means that subsequent units may become substantially cheaper though the experience gained. Orders achieved in China, Finland and France are the start of overcoming this. Some creative public assistance with initial units could be justified in a number of ways, related to energy security of supply concerns, diversity of power sources and avoidance of carbon emissions. They could take the form of loan guarantees, accelerated depreciation allowances and investment or production tax credits. Those opposed to nuclear would see these as subsidies, but they are familiar measures taken in favour of certain power generating options (such as wind power today) and other industrial objectives in many countries. It would be expected that these would last for only the first half dozen units with the expectation that following units would have lower capital costs with shorter construction times and be fully economic.

There are clearly substantial challenges in financing investments of $3 billion and above when the only revenue comes from electricity sales in markets that may be unsteady. In liberalised markets where prices are set hourly by marginal cost offers, risks of new investment are substantial, yet longer-term needs of customers have to be covered in some way. However, as markets mature, longer-term commitments may be easier to make.

One approach is for major power customers to take equity stakes in nuclear plants. This is the model adopted in Finland where the owners of the fifth nuclear plant, ordered in late 2003, will take most of the power produced. Assuming the reactor is built to schedule and operates as expected, the owners should have long-term power supplied at 2.5 euro cents per kWh, including all capital and interest repayments.

In conclusion, currently operating nuclear power plants have clearly demonstrated their strong economic credentials. Once nuclear plants are up and running, they should make excellent profits for their owners and have long operating lives. Establishing the sound economics of new nuclear plants in the minds of potential investors remains more challenging. The various key parameters in the equation are well-understood and set out in the many available studies. In particular, capital costs must be kept as low as possible and financing secured at reasonable costs of capital. Where this is achieved, for example with the fifth reactor in Finland, the economic case can be very strong. If the increase in fossil fuel prices is sustained for a considerable period, nuclear competitiveness will receive a notable boost, but the industry is wise to concentrate on smoothing the regulatory path and obtaining cheaper financing. In countries where the benefits of nuclear power are well-recognised, support for the initial units in a wider nuclear building programme may be forthcoming.

New nuclear plants should eventually be regarded as good, conservative long-term investment prospects. Once the initial significant capital cost burden is overcome, they can offer electricity at predictable low and stable costs for up to 60 years of operating life. Investment in nuclear should therefore be attractive to industrial companies who require significant amounts of cheap power for their operations, but also fund managers in the financial sector who are managing pension and life assurance funds.

Liberalised power markets: how can nuclear fit?

Electrical power generation, including nuclear, was largely developed by public bodies in a regulatory environment that permitted long-term investment but passed on the full cost to customers. Even if generation was not in public ownership, power utilities could recover most of their costs through the electricity rate-setting mechanism without having to worry too much about competition. Nuclear plants are expensive to build and can only recover their costs over a long time period, so clearly benefited from this environment.

National energy planning was also in vogue at this time and securing a significant nuclear sector, perhaps to economise on fossil fuel imports and to increase energy security, was followed by major countries such as France and Japan. Indeed, it can be argued that many nuclear plants were primarily built for national security of supply reasons, although the promise of cheap electricity with a stable cost base was clearly very important too. Even today, reducing the dependence on imported fossil fuels with uncertain price prospects remains important in countries without substantial domestic oil, gas and coal reserves. The expected long-term stability of costs was also an important consideration in favour of nuclear and remains a strong argument today.

The gradual liberalisation of power markets throughout the world has posed many challenges to the supply sector and commentators initially suggested that the impact would be strongly adverse for nuclear power. For example, a few years ago it was widely believed that perhaps half of the 100 commercial nuclear reactors in the United States would be victims of liberalisation and could close down within ten years. The burden of 'stranded' costs (*i.e.* those attributable to delayed plant construction and unlikely ever to be recovered by subsequent profits) and the possibility of future operating costs lying above the expected electricity price would mean that shutting down could be the best option.

In practice, this hasn't happened – indeed it can be argued that most existing nuclear plants throughout the world have thrived in the era of liberalisation. Stranded costs haven't really been an issue, as so long as a plant is operating at marginal costs below the market electricity price, the best option is to carry on running. Competition between generating assets was the spur to improve performance that was needed by many nuclear plants and the industry has clearly responded. The evidence is most dramatic in the United States where average capacity factors have risen over the past ten years from the mid 70s to 90% and marginal operating costs are now lower than for coal and gas plants. It is now believed that reactors will run for their entire licensed periods and many plants are now applying for and gaining 20-year extensions to these. It has been worthwhile to invest in necessary major plant refurbishment, such as new steam generators, while many reactors have also achieved significant power uprates.

Elsewhere, the picture is much the same and it is now generally agreed that existing nuclear plants have little to fear from market liberalisation. Once a nuclear plant is up and running, marginal operating costs should be relatively low and able to beat the market price. Power utilities in Germany and Scandinavia, for example, have found their nuclear plants to be very profitable in the new environment. Pressure to close them has therefore been political, not economic. The situation in the United Kingdom has been more complex. In the early days of electricity liberalisation, the nuclear plants were very profitable but the dramatic fall in the power price following the revisions to the electricity market trading mechanisms changed this. These difficulties were shared by other generation sources, but the age and technology of the British reactors have recently made it difficult to achieve the improved plant performances seen in other countries.

Nevertheless, the economic case for running nuclear plants for a long time within liberalised power markets is a strong one. Possibly 350 of the existing 439 reactors currently in operation throughout the world will still be running in 25 years time. Those that close will be older, smaller units and therefore more difficult to justify spending money on, or those which are victims of politics.

Turning to the prospects for new nuclear plant construction, it is certainly true that market liberalisation poses a significant challenge. Many potential investors require a quick return on their money and the financial profile of a nuclear project does not fit in with this, even if plants can be built in 4-5 years and within cost. The construction costs of a combined cycle gas turbine (CCGT) plant are roughly one third of those of a nuclear plant of similar generating capacity, which is a substantial barrier to overcome. The comparative advantage of nuclear plants is their proven low and stable fuel costs, whereas with a gas plant, an investor is mainly running the risk of escalating fuel prices, which are the most significant cost item over the life of the plant. However, so long as relatively low gas prices can be predicted for the first 5-10 years, a CCGT can pay for itself in this relatively short time period. An investment in nuclear, by contrast, needs a much longer time period to receive the benefit of the low fuel costs.

If we look today where new nuclear plants are planned or already under construction, there are two major areas. Firstly Asia, where both China and India have ambitious nuclear programmes, to add to the existing, more mature plans in Japan and South Korea. China and India see nuclear as important in satisfying the electricity needs of their huge populations, at a time when economic growth is very rapid. These plans are very much directed by state planners but both are confident that reactors can be built on time and at economic cost. Both countries are heavily reliant on coal for power generation and nuclear's zero carbon emissions are another sound reason for including it as part of their plans. The other area where new nuclear is still very much alive is the former Soviet Union. The task in recent years has been to complete those reactors already under construction when the Soviet Union collapsed but attention is now returning to the ambitious plans both Russia and Ukraine had in the past. As the financial situation improves, the strong nuclear expertise in these countries will encourage the completion of further reactors. Again, the electricity supply system is very much under state control, but this allows a long-term perspective to be taken.

The case not fitting in with this pattern is the fifth reactor that has just been ordered in Finland. Here the power market is liberalised and government intervention is very limited. How can this environmentally-friendly European country order a new reactor when others, such as Germany and Sweden, are planning to close nuclear plants which operate safely and economically? The answer is essentially in two parts. Firstly, Finland's experience with its existing four nuclear reactors has been very favourable. They have operated very well and at low marginal costs which have meant good profits for their owners. Secondly, and most importantly, the companies taking the power once generated will own the new Finnish reactor. Finnish industry is electricity-intensive and the heavy users require the security of long-term baseload power at low and secure prices. The performance of the existing reactors suggests that this is achievable so, after a detailed study of the alternatives, they have chosen another nuclear plant.

The question is whether the Finnish experience can spread to other liberalised power markets in Europe and North America. Probably the key country is the United States, where there will be a major requirement for new baseload power plants over the next 25 years, to satisfy rising demand but also to replace old fossil fuel-powered plants which are nearing the end of their lives. The US government is certainly supportive of nuclear and developed the *Nuclear Power 2010* programme with the industry, aiming to get a new reactor underway by 2010. The most important requirement is to bring new reactors into operation rather more quickly than the bad experience of the past. The legislative side is now rather easier in principle, with early site permits (ESPs) for extra reactors on existing sites and combined construction and operating licences (COLs). The difficulty, however, is finding the first power company to take the leap and order a new reactor. It is most likely to be consortium of owners, but the past experience of nuclear construction in the United States and the attention a new order will gather is making both plant operators and their financiers somewhat cautious.

Although the fifth Finnish reactor has helped in making the economic case for new nuclear build in liberalised markets, the delays already experienced have been a significant blow. In China, however, the recent experience is more favourable with most reactors brought online both on schedule and within projected costs. But the industry still needs to find ways to cut costs by achieving full economies of scale, simplifying designs and modularising construction where possible. There are probably too many companies in the reactor design and construction business at present and, like the aircraft manufacturing sector, should move towards producing a limited number of standardised designs in great numbers. Lots of different designs in one country, as the United States has at present, is unlikely to meet the economic criterion.

The other area the industry can work on is to identify opportunities in financing new plants. Major industrial power users could have a nearby nuclear plant satisfying their requirement and any surplus may be sold on the open market. This should remove some of the risk of investing in a nuclear plant, as the major part of the revenue will be guaranteed. The alternative approach is to encourage the development of longer-term power markets where companies can buy and sell contracts covering greater time periods. Certainly there is scope for financial innovation as liberalised power markets are still very young. Moreover, there are significant doubts as to whether they can guarantee an appropriate level of investment in new capacity (no matter whether it is coal, gas, nuclear or whatever) in order to create satisfactory safety margins to cover peak demands from customers. The state used to take care of this, but it is now not clear that the market will allow this.

In conclusion, we can say that liberalisation has been generally good for existing nuclear plants but it is still unclear whether new plants can prosper in this environment. The industry has certainly got much to do in order to improve its prospects but is also reliant on how liberalised markets themselves develop.

Financing: a barrier to new nuclear build?

Over the past couple of years, there have been a number of studies showing that new nuclear power plants can definitely be economic propositions, when compared with the alternative of building gas- or coal-powered generating capacity of similar magnitude. The most important factors in these assessments are now well-understood –

particularly the construction cost of the nuclear plants, the price of gas and the rate of interest assumed in the plant financing. Yet nobody is rushing to build new nuclear plants in the Western world, when the studies show that investors could potentially make a lot of money out of doing so. Why is this?

The entrenched and strong political opposition to nuclear in many countries is undoubtedly a factor yet in the United States, clearly the key market for new nuclear build, the government has now come out strongly in support, offering various financial incentives in the Energy Policy Act of 2005. But the power companies are still sitting on the sidelines, with their senior executives very cautious. It is clear that their major fear is the reaction of financial markets to any further engagement in nuclear power. Their credit ratings are likely to take a hit while the impact on their earnings per share of a major investment project, with no financial return for several years, must be carefully considered.

It is clear that financing new nuclear build in the financial markets will prove very challenging. Nuclear has a bad reputation there, with memories of the cost over-runs of plants in the 1970s and 1980s still strong, with the attendant power utility 'stranded costs', *i.e.* those investment costs never likely to be repaid by future electricity sales. There is a general perception that nuclear has not yet cleaned up its act – the problems at British Energy (originally caused by power market reforms but now spread to plant engineering issues) have caused further image damage to the industry. Financiers have been loath to invest equity in new nuclear build, while loan finance is likely to come only with a hefty risk premium, such that the economics of new plants may not work out. In general, financing needs to be available at under 10% per annum to make new nuclear build work economically.

The key issue in plant financing is the clear identification and management of the risks. How much can the private sector possibly bear and which risks need to be left in the government domain? The balance here is very contentious. The industry's public image has suffered badly owing to the accusation that it has always been heavily-subsidised and cannot pay its own way. It is clear that nuclear contains enough 'special features' to make it unlikely that it can ever break completely free of state involvement in its affairs, but the industry now seeks to achieve new plant orders on a fully-commercial basis. Yet, at the same time, as energy is a strategic commodity, governments are seen as retaining a vital obligation to help manage and mitigate the obvious risks.

Where do these boundaries lie? From the industry's viewpoint, it has a number of requirements of governments, which it seems reasonable for them to fulfil. First amongst these is a sound regulatory system, which allows reactor design certification to take place quickly, then allows construction and operating licences to follow without unreasonable delays. The local planning process is also important, as endless appeals against decisions can destroy nuclear plant economics, when the key requirement is to get a plant built and earning good revenue as soon as possible. These were the key areas where nuclear build got tripped up in the United States before, so there are currently important efforts being made to overcome them with a new generation of plants. Design certification of new reactor designs such as the AP1000 has gone quite smoothly, while the early site permit (ESP) and combined construction and operating licence (COL) processes are proceeding apace. This will do a lot to assuage investor concerns and should be replicated in other countries too.

There are, however, other important areas where governments should be expected to act. There are issues such as nuclear safety, plant security and nuclear liability provisions where clear guidance, with consistent rules and regulations, are a reasonable expectation. Nuclear non-proliferation has now also become a major issue – yet the industry has little to fear if governments do their proper job of sorting out a workable international regime, which satisfies reasonable public concerns.

Perhaps the biggest requirement of governments, however, is deciding on (and then delivering) a workable national radioactive waste management policy. Without being able to demonstrate a practical solution to the waste issue, it will be difficult to build new nuclear plants. In the United States, the industry has rather put its eggs all in one basket, in the shape of the Yucca Mountain repository, to the extent that further delays could conceivably constrain further new nuclear build. The industry quite rightly insists that waste management is an area where they have offered a range of feasible technical solutions, but governments have failed them by lack of action. The related issue that comes up, particularly in the United Kingdom, is that of eventual plant decommissioning. It is hard to explain to even experienced financial analysts that this is really irrelevant as far as pure new reactor economics goes – decommissioning is so far in the future that the necessary payments to a fund for covering it are relatively small and don't really bear on new build economics. However, the industry still needs the state to set the rules here – some would argue that it must also act as the ultimate guarantor as a plant owner could conceivably find a way of walking away from its future obligations, even with an adequate decommissioning fund in place.

Finally, it is a reasonable requirement of governments that they should create power markets where different technologies can compete on a level playing field and where long-term investment in capacity is incentivised. The state's ultimate responsibility is clearly to prevent the lights going out by ensuring a reasonable capacity margin, but, as the revenue from billions of kWh of electricity is the only outside funding coming into the nuclear industry (a point often forgotten by those engaged in uranium production and other fuel cycle activities), the shape of power markets, the contracting and other market mechanisms, are very important. Investors have to take major risks with selling nuclear power at good prices for many years in the future, to recoup the heavy initial plant investment costs.

These requirements of the state are almost essential preconditions for cutting the risks of new nuclear build to manageable levels. That they have so often been absent in the past explains many of the subsequent problems the industry has even today. There are, however, further risks that a combination of the financiers and the plant owners will have to share between them, relating to the lifecycle of a nuclear plant.

Clearly construction delays, caused by technical or labour problems, have to be avoided. It has been demonstrated in Asia that new plants can be operational only four years after initial concrete is poured and this has to be replicated everywhere else. The sooner electricity revenues start flowing to repay the interest and capital costs accumulated since construction began, much the better. Once the plant is online, there are then a range of risks relating to its operations, including keeping its capacity factor at a high level by careful outage management, avoiding other time offline and generally keeping the plant economics in a sound shape. There are also risks relating to further capital expenditure. Key plant components, such as steam generators, will eventually have to be replaced but such costs have to be minimised as they are important to overall plant economics. Yet looking ahead, they can nearly always be justified by the advantages of receiving many further years of strong revenue from electricity sales.

Adhering to all the rules set by governments on safety, plant security and non-proliferation is obviously important, but nuclear plants around the world have generally shown that they are very good at this. Once left to get on with the job of producing lots of electricity cheaply and safely, risks should be quite low. Yet when operators are careless, as has been the case in Japan, regulators will step in and close plants for extended periods until they are satisfied. The availability of sufficient trained personnel to staff new build is an issue that often comes up in discussions but is one that is soluble by a combination of the initiatives already underway in industry and education – perhaps, most easily, by paying plant operators rather better and increasing the prestige of working in what is sometimes regarded as a graveyard sector of the economy.

Looking at the whole picture, we may conclude that most of the risks which investors fear are manageable if governments set out a proper framework for the industry. This doesn't include subsidising nuclear in a financial sense, only setting out a framework in which it should be allowed to operate, ensuring public safety is maintained and that new plans are subject to reasonable scrutiny by those directly affected. The risks of building and operating the plants seem quite minor in comparison as the industry can claim a good record when it is allowed to get on without expensive interference from outside.

Understanding all the risks and allocating them to their correct location is clearly something that financiers find very difficult. Nuclear projects are very complex when compared with building a gas-powered plant, bringing forward a complete new range of issues that they are neither familiar nor comfortable with. In addition, it has now been recognised that nuclear can contribute to national greenhouse gas and security of energy supply objectives, which may persuade governments to guarantee additional revenue which need not be seen as subsidies – rather, a justified reward for offering something which other power generating modes do not.

These risks (and rewards) of nuclear investment are something the industry needs to explain much better to the financial community. Given the ability of the financial sector for innovation, it is likely that other ideas will be introduced to cope with the unpredictable developments of electricity markets. Consortia may be put together which have the confidence to invest in nuclear. Investors such as pension and life assurance funds should find nuclear plants attractive as their returns should be stable and predictable over long periods. It may also be possible to introduce long-term forward contracts for power sales, to allow major capacity additions to take place.

Yet the industry has a big selling job to do, in order to encourage them to put their minds to any innovative financing mechanisms they can come up with. It is not yet certain that unregulated merchant generating plants will everywhere be the favoured solution in power markets and regulated utilities, public private partnerships and power users investing directly in plants (as is the case in the fifth Finnish unit) remain open possibilities. Similar opportunities with major power users should be explored by the nuclear industry. The need is for a major part of the electricity to be effectively pre-sold, with the remainder sold on the spot market or through shorter-term contracts. The financial sector can be incredibly creative once it is convinced that there is good money to be made in something – it is now at the stage of just beginning to see this with nuclear, but feels weighed down by the huge number of potential risks. It must be allowed to see more clearly – definitely a case of letting it see the good quality wood amongst all the leafy trees.

New nuclear build: is there sufficient supply capability?

It is clear that the envisaged new nuclear build programme in the UK will be almost like establishing a new industry. Despite significant experience of nuclear in the past, indeed as a leader in nuclear technology from the 1950s onwards, the UK has not had a substantial nuclear reactor construction programme for many years. It is therefore reasonable to question whether there is now the capability to supply the plant and equipment (either from local suppliers or from the international market), to carry out the major civil engineering and construction works which will be required, and finally to have adequate programme management and technical support. Within each of these areas there is an underlying concern that the people and skills may no longer be available or be stretched by their involvement in other competing activities. For example, within the UK nuclear industry itself, the change in focus towards plant decommissioning and the cleanup of old sites is tying up a lot of experienced staff. At any one time there are also other major infrastructure projects underway, which can take up a substantial share of the available national resources.

It is interesting to make a comparison with South Korea. The nuclear programme there has allowed the development of a local reactor design and construction infrastructure in the form of companies such as KOPEC and Doosan, which has had a steady stream of work for many years. This has followed consistency of public policy on nuclear and would allow an acceleration of new reactor construction to take place relatively easily. In the UK and the West, however, the tap got turned off during the 1980s and the ability to respond quickly to the needs of a new programme has suffered. The question is: how badly?

Some useful analysis of the issues underlying this question, at least for the UK, is provided by two complementary reports, both looking at the national capability to support a new nuclear build programme. These reports are: *The UK capability to deliver a new nuclear build programme*, Nuclear Industry Association (2006); and *An evaluation of the capability and capacity of the UK and global supply chains to support a new nuclear build programme in the UK*, IBM Business Consulting Services (2006).

The Nuclear Industry Association (NIA) used a working group of its members to undertake its study, focused very much on demonstrating that UK companies still have the ability to support a programme of ten new reactors over the next 15-20 years. IBM Business Consulting Services carried out a telephone interview programme with UK and global suppliers, highlighting where overseas sourcing will likely be necessary. It concentrates more on the plant and equipment area, whereas the NIA study is far more comprehensive, including detailed evaluations of the civil engineering and project management areas too.

The NIA believes a typical nuclear power plant can be divided by value into plant and equipment (55%), civil engineering and construction (30%) and project management and support activities (15%).

Within the plant and equipment area, a subdivision can be made on the nuclear part of the plant (the nuclear island) and secondly the non-nuclear part (which would be similar for any thermal electricity generation station) – the split on construction costs is likely to be approximately 50:50. At the time of the last nuclear plant to be built in the UK, Sizewell B, almost all plant and equipment could theoretically have been supplied by UK companies, the major exception being the reactor pressure vessel and other large forgings. Competitive tendering led to additional contracts being won by overseas companies, but a significant level of local nuclear engineering, manufacturing and site installation capability still existed.

Since then, it's clear that there has been a dramatic downsizing and redirection of manufacturing capability in the UK for large power stations of all types, not just nuclear, essentially because of little market demand. This applies particularly to large turbines and generators and their associated switch and control gear. The supply of large forgings is a critical area, particularly the head forgings for reactor pressure vessels and steam generators. Only a few overseas companies supply these and if there is a general revival of nuclear build worldwide, lead times on supply could steadily increase. On the other hand, if there is confidence that there will be a steady stream of new orders for many years, investment in new facilities by these and possibly some new suppliers will likely take place. The importance of this area is underlined by the acquisition by Areva of Sfarsteel, an integrated producer of steel forgings including the historic Le Creusot forge, which makes the very large components of the type used in nuclear power plants.

Overall, it is believed that about 50% of the necessary plant and equipment could be delivered from current UK facilities and resources, which may be increased to 70% with further investment by local companies, assuming they are given the confidence to do so by the assurance of a substantial multi-plant programme.

Most of the civil engineering associated with new nuclear build is similar to other major civil engineering projects. Although the magnitude of the programme envisaged is certainly significant, it would be no larger than other major infrastructure projects, such as oil and gas terminals and the 2012 Olympics site in London. A relatively small proportion of UK construction materials would be required, less than 1% of annual cement and aggregate output and less than 4% of structural steel production. A modularised approach to construction may well be employed, reducing onsite labour requirements at relatively remote locations, but requiring the ability to handle the large modules. In terms of timing, other major infrastructure projects in the UK are timed for completion prior to the start of a nuclear new build programme, so that it would provide continuity of work rather than overstretching.

In the programme management and technical support area, it is possible that a grouping of companies will provide the necessary resources, rather than a single UK or overseas company, given the need for a strong and experienced leader with an international reputation. Similar to the civil engineering area, the resource demands will be small compared with the overall UK capability, but there may be some constraints in areas such as reactor safety and licensing, where many of the current experienced personnel will be approaching retirement age over the next five to ten years. It will be possible, however, given the phased and extended nature of a new build programme, to implement training programmes to provide a new generation of technical specialists to fill the gap and provide continuity of support.

The overall position with regard to skilled staff is one frequently mentioned as a possible constraint on a new build programme, applying in many areas. In reality, however, even a ten-reactor programme will be spread over many years and observations such as that many UK reactor operators are close to retirement are not particularly relevant. There will be time to train a new generation of operators and it should be an attractive career choice, given that new reactors will likely run for 40-60 years. The new build programme will also employ an international reactor system, already designed and licensed in its country of origin and probably already constructed at another location. This contrasts with previous UK experience with the Magnox and AGR programmes, but also with Sizewell B, which started as a standard PWR but was extensively modified. Without the need any longer to develop the rector designs, skill requirements can be focused on selecting, licensing and constructing a standard reactor system within the UK regulatory framework.

It is therefore clear that the UK supply chain is capable of delivering most of a new nuclear build programme but will require support from the wider global chain in a few key areas. There is some risk that there will be capacity constraints worldwide imposed by similar new build programmes in other countries but what is really needed to address this is a higher degree of certainty. Companies will be very happy to invest in new facilities and staff as soon as they are assured that there will be a steady flow of demand for the foreseeable future. At the moment, there is a lot of fine talk but only a limited amount of progress towards this. Yet this is only to be expected as a return to new nuclear build in the Western world represents a major break with the recent past. There may be some shortages of capacity in the early stages but, given the timescales of nuclear projects, it is not complacent to expect markets eventually to react and bring forward the required staff, materials, components and services by the time they are required.

Above all else, governments have to demonstrate the political will to support new build, develop regulatory systems to ensure that reactors can be approved and built in a reasonable time and also finalise the necessary public policies on waste management and decommissioning. The industry cannot expect guarantees that things will never change, but needs some general degree of political consensus to invest – if it's going to be used as a political football as in the past, there is little prospect of new plants.

New nuclear plants: what are the real issues?

Nuclear conferences are full of discussion about the renaissance in the industry. The evidence for this comes from several quarters – from the much improved performance of current reactors which is leading to power uprates and licence extensions, from the increased interest in building new nuclear plants in the US and UK as well as in China and India and also from improvements in expressed public support for nuclear in many countries. Certainly the industry is able to walk much taller than in the recent past and is now receiving interest from respectable bodies formally, at best, agnostic. For example, the International Energy Agency, long seen as a bastion of fossil fuel dominated thinking, included a chapter on nuclear in the 2006 edition of the *World Energy Outlook*. Some sceptics claim that the only reason for the increased attention is nuclear's possible role in greenhouse gas abatement, but the energy security and the economic arguments are also slowly beginning to win over. The economic advantages are undoubtedly the most difficult to put across, as there is still widespread scepticism about the viability of new nuclear build, but they are arguably the most compelling – cheap and reliable power is a great selling point.

There is certainly, however, a risk that the industry will begin to see the battle as already won. It is indeed tempting to imagine the industry's profile in terms of a sharp rise in fortunes in the 1950s until the 1980s, then a 25-year period where things flattened out, only to be followed by a resumption of the strong upward path from today onwards. Mentally picturing a graph, we can envisage a strongly rising trend in the period from 1950 to 2050, but with a flattened kink in the middle. At the time, the kink may have seemed like maturity for the industry, to be followed by a long decline as reactors gradually shut down without much new build as replacement capacity (and certainly little as incremental capacity).

Indeed, up until relatively recently, this was how nearly everyone, including many people in the industry, felt things would most likely develop. But now more and more people have the vision of resumed growth, maybe rather slowly at the beginning but accelerating as time goes on. For example, the World Nuclear Association's (WNA's) upper scenario for nuclear generating capacity shows a doubling on a worldwide basis by 2030. Yet some other pundits have observed that this may now even be a rather conservative vision, with even faster growth possible from around 2020 onwards.

But is this still just a wild dream? The essential element is surely obtaining lots of orders for new reactors and as soon as possible too. Keeping the existing reactors going for longer has been a tremendous achievement, but now the industry has to prove that it can make a further huge dent in carbon emissions, contribute to enhanced energy security, while supplying billions of cheap kilowatt hours for the masses. There is plenty of talk of new reactors – in the US, now the UK too, China, India, Russia and also many other countries, some of which currently don't have nuclear power (such as Indonesia and Vietnam). Yet talk can be very cheap and there is always a risk that it will amount to little than more than just that. People in the financial sector say that at any one time, they are looking at huge numbers of prospects for new investment, but nearly all fall by the wayside – could new nuclear build meet a similar fate? What is needed to prevent this?

Two important things are already in place – sound potential economics and better public acceptance. There have been a number of recent studies showing that new nuclear power plants can definitely be economic propositions, when compared with the alternative of building gas- or coal-powered generating capacity of similar magnitude. The most important factors in these assessments are now well-understood – particularly the construction cost of the nuclear plants, the price of gas and the rate of interest assumed in the plant financing. Yet nobody is yet rushing to build new nuclear plants in the Western world, when the studies show that investors could potentially make a lot of money out of doing so.

Secondly, it is rapidly becoming apparent that the arguments of the anti-nukes are sounding tired and worn. It maybe slightly premature to say that the long intellectual battle fought by the industry against its opponents has been won, but it's increasingly looking that way. Public opinion is moving to the industry's side as people are becoming comfortable with the solutions proposed to deal with the challenges the industry faces on safety, proliferation of nuclear weapons, waste management and plant decommissioning. Although local people must be carefully involved in plans to build new reactors, it seems that large-scale national campaigns against new build will now be much harder to sustain. This, of course, depends on the continuation of the industry's excellent safety record and progress in satisfying reasonable demands to activate solutions to nuclear wastes, which is probably the most difficult remaining issue in obtaining full public acceptance.

There are then a couple of issues that have been posed as barriers to new build, which are really not so difficult when they are considered more closely – namely financing and possible shortages of materials, plant components and staff.

There has been a lot of talk about the difficulties of financing new nuclear plants. Yet assuming the economic case is strong, finance is unlikely to be a barrier, although it is clear that financing new build will prove challenging. Most equity finance may have to come from existing power market companies, and loan finance is likely to come only with a risk premium. Yet there is a huge wall of money available internationally for new investments of all kinds – the difficulty the money people have is in locating sound prospects. The key issue is the clear identification and management of the risks. Structuring a project in an appropriate way is vital – in the case of nuclear, in particular how much of the risk is to be taken by the private sector and which will be unavoidably left in the government domain. The financial sector can be incredibly creative once it is convinced that there is good money to be made in something – they are now at the stage of beginning to see this with nuclear, but need help with understanding the nature of the potential risks. But finance is essentially an output, rather than an input, to a good project – there is no magic wand that can be waved to rescue what is fundamentally a bad project anyway. But once a good plan has been devised, the financiers will do the necessary.

It is clear that a new nuclear build programme in the UK, for example, will almost like be establishing a new industry. Despite significant experience of nuclear in the past, there are now at least potential shortages of men and materials for a new programme, particularly if other countries are building many reactors at the same time. This is undoubtedly a global issue, but the timescales involved should be sufficient to allow solutions to be found. There are already many programmes in place to remedy the deficiencies observed in nuclear education and training, but nobody is going to invest in capacity to produce the necessary plant components unless the orders are certain and likely to be repeated. Yet given the timescales of nuclear projects, it is not complacent to expect markets to react and bring forward the required staff, materials, components and services by the time they are required.

If the above factors should not now be big barriers to new build, where are the potential weak points? These lie in two main areas – firstly with national governments and secondly with the companies which should lead the new projects. The real challenge to a nuclear revival will come from either a lack of political will or a lack of guts from those charged with making decisions on new generating capacity. We can see each of these by referring to the UK and the US, in both of which a new nuclear build revival is crucial as a lead to the remainder of Europe (UK) and the rest of the world (US).

In the UK, the government has to demonstrate the political will to support new build. It must develop the regulatory system to ensure that reactors can be approved and built in a reasonable time, finalise policies on waste management and decommissioning (so the plant owners know exactly the extent of the financial provisions they must make) and satisfy the requirements on plant security and nuclear liability. The United States has got rather further ahead in most of these, so the issue there is more a case of who is ready to invest.

Nuclear power plants are highly complex projects and it is almost understandable if a power utility chief executive holds up his hands in horror at the prospect and says: "Let's build a gas or coal plant instead." This may be a much safer business decision, so some courage is needed, at least for the first investor to commit. It is clear that the major fear of big power companies is the reaction of financial markets to any further engagement in nuclear power. Their credit ratings are likely to take a hit while the impact on their earnings per share of a major investment project, with no financial return for several years, must be carefully considered. It may need some initial government subsidies to encourage them, such as the loan guarantees and production credits already proposed, but these will hopefully be sufficient to get the ball rolling.

The industry cannot expect guarantees that policies will never change, but needs some general degree of political consensus to invest – if it's going to be used as a political football as in the past, there is little prospect of new plants. Brave investors, with strong belief in the long-term benefits of nuclear and the willingness to put their money where their mouth is, are crucial as well. The nuclear sector should then be able to stand on its own feet and show that it can indeed generate a huge quantity of power economically and environmentally-soundly, while contributing to national and regional energy security of supply.

3. PUBLIC ACCEPTANCE

Introduction The word 'nuclear' tends to ring alarm bells with almost everyone. Owing to its popular association with 'bomb', 'war' and 'disaster', it is not surprising that nuclear power has an image problem to overcome. The commercial industry, as it stands today, has obvious roots in the weapons programmes of the past, but struggles to distance itself from the adverse image that this has created. Detailed reporting of even minor incidents at nuclear facilities, on the basis that they are only a hair's breadth from involving massive death and destruction, has always been expected. This is sufficient to keep active anti-nuclear movements alive in most countries around the world, even as memories of the Chernobyl accident over 20 years ago have faded.

Companies involved in nuclear power have put a huge amount of effort into improving both their own image and that of the technology itself. This has been necessary partly because the early days of the industry were marked by a degree of arrogance in public communications which did nobody any favours and which would be unacceptable today. The poor public image served to impose additional costs on the industry through delays in completing projects and the need to comply with the imposition of highly prescriptive regulatory regimes. Achieving a better public image therefore makes a lot of economic sense and is worthy of the time devoted to it by many people in the industry. Experience has shown that this is best accomplished at the local level, by letting people visit nuclear facilities and ask lots of questions. Nuclear power plants now achieve high approval ratings in nearby areas, and not just because of the jobs and income they create. Indeed, in Sweden different towns were competing to be the location of a new waste facility and in China, it is local governments who have been pushing to have new nuclear plants rather than central government planners.

Nuclear has often been depicted by its detractors as an evil technology imposed by wicked central government planners and multinational companies on unwilling local people who happen to live near suitable sites, perhaps by a lake or the sea. If the critics can be marginalised as a self-interested largely urban minority with naïve designs on a return to a simplified pre-industrial past, the advocates will have succeeded. There is now ample evidence that this is, indeed, slowly happening.

The roots of opposition to anything to do with nuclear run very deep and are an essential element of the environmental movement, which has found the industry a relatively easy target. It has been helped by an initial poor response from its advocates and some structural weaknesses in the industry itself. The 'nuclear lobby' is often depicted as being very powerful, but in essence it is relatively weaker than that of other prominent industrial sectors, such as the oil, chemicals and auto industries. One argument that has, however, begun to come across is that, contrary to the view that nuclear imposes both explicit and latent costs upon society, the truth is in fact the opposite. Its 'external costs', to use a technical term, are essentially largely 'internalised' whereas other sectors are free to pollute without penalty. This is where the climate change debate has greatly benefited nuclear, as it has focused attention on the sins of the fossil fuel sector, not just in terms of emissions of greenhouse gases but also inner-city pollution and the wider accident rates in energy production.

Potential proliferation of nuclear weapons is one risk that those opposed to nuclear highlight as a reason for shutting the industry down. It can be seen, however, that there are substantial mechanisms in place, which should allow people to sleep more easily in their beds at night. Although some nuclear advocates have enjoyed poking fun at some of the more extreme claims of the renewables lobby, they generally argue today that both can work side-by-side in achieving a more sustainable energy strategy. If climate change is such a problem, surely we should not be ruling out any low carbon technologies that can help. Unfortunately, those advocating wind, solar and tidal energy solutions are amongst those most strongly opposed to nuclear and it has been difficult to build bridges with them. There is some concern that major new nuclear programmes will 'crowd out' investment in renewable forms of energy, but this needn't necessarily be the case. Attention paid to getting the true facts across about nuclear to a wide audience can only get the industry so far when people's values are so set another way. Third party advocates have been very helpful, but there are those who will never be swayed, as nuclear embodies all they hate about modern society.

On the more respectable wing, there are some more intellectual and possibly persuasive criticisms of nuclear than is usual from the likes of Greenpeace and Friends of the Earth, whose arguments are looking increasingly threadbare. But again, the differences tend to boil down to interpretations of facts and perceptions of risks, rather than anything which can easily be overcome. The best solution for the industry, therefore, is probably to carry on doing what it does best, which is running the facilities well, both safely and economically, and showing the public what it is doing without hectoring them. PR campaigns are likely to be self-defeating – perhaps the industry should worry a little less about public acceptance, now that the battle appears to being slowly won, and concentrate on those aspects over which it has much better control.

Nuclear's adverse public image: where are the roots?

Nuclear industry people often rest under the illusion that their business is the only one under attack by strong opponents, engendering a feeling of isolation and supreme defensiveness. Far from it – these days no industrial sector gets an easy ride with public opinion. Under the umbrella of corporate social responsibility (CSR), all industrial sectors must justify their activities in terms of their environmental and social impact. The presumption today is essentially 'guilty until proven innocent'.

Nuclear expanded in the 1950s and '60s when the world was very different. This was the era of technocratic decision-making, where a brave new world was to be created, led by the application of science in new technologies. This lasted until the mid 1960s, by when a lot of scepticism had crept in and industries began to have to justify themselves. Nuclear is not unique in this regard, but it is arguable that it provides a particularly strong example of how the world has changed. But why does it generally have such a poor image with the general public?

To start with, we have to accept that those passionately opposed to nuclear have played their cards very well. The environmental movement picked well when it selected nuclear as the key issue around which all their camp followers can unite. Rather like Christians having to believe in God, being anti-nuclear is a prerequisite for being a true Green.

There were undoubtedly alternative possible totem poles, notably attacking private car ownership. But it was recognised that this is one of those things (rather like extra-marital sex and also price inflation), which are universally publicly condemned, but greatly enjoyed in private by many. So best not to make an issue about automania and its environmental consequences – better to pick on nuclear.

Many of these strong opponents see nuclear as the thin edge of the wedge – they would seek to shut down the next industrial sector if successful with nuclear. Indeed, many appear opposed to modern industrial society, economic growth and globalisation. Given the evident weakness of their case and the few people who strongly support them, they have achieved a great deal.

One particular problem the industry faces is that it is readily identified as a sunset sector. Without a major plant-building programme, backed by an irrefutable economic case, the industry faces its critics with a hand tied behind its back. This is arguably its biggest public perception problem. In the Western world at least, the talk is more of plant decommissioning, much of it reflecting the burdens of the military nuclear legacy. The only significant reactor construction programmes are taking place in countries which critics may isolate as 'special cases'. The industry needs to reverse this trend somehow – in other words to get those countries not involved in nuclear expansion labelled as special. This will require the development of a superior economic case, as for example in Finland, the one shining example in today's picture of general gloom.

Another significant problem from which nuclear suffers is the lack of strong industrial companies prepared to stand up and champion it unequivocally. In other major industries, public support has been gained by feeding on the strong image of the biggest corporate players. Take the oil industry as an example. Its activities are highly questionable from environmental and sustainability angles, yet it generally gets an easy ride from the public. Some of this undoubtedly flows from popular worship of the private car, but the strong public image of leading companies such as BP, Shell, Total and Exxon certainly helps the overall industry perception.

Part of this problem is that the nuclear industry isn't really an industry at all, rather a set of separate businesses participating in various parts of the nuclear fuel cycle. The sole conglomerate, covering several areas, is Areva. BNFL was formerly in that position, but has now been broken up. Both of these have made major efforts to strengthen their public images, with some success, but this has so far been insufficient to greatly improve the public image of their industry. Other large companies interested in nuclear are attached to alternative technologies and therefore won't stand up to be counted. For example, the electricity generators will use nuclear, coal, oil or gas, with no strong attachment to any of these, while major industry suppliers such as Rio Tinto are heavily involved in coal or other energy sources. This has always been a significant problem for the World Nuclear Association (WNA) in encouraging a more positive image of the industry. There are many worthy individuals amongst its member organisations, dedicated to the future of nuclear, but they are insufficient to carry great weight within their own companies, which are successfully involved in many other areas. It has made it particularly difficult for WNA to get strong advocates from amongst the key public affairs spokespeople of these companies.

Another issue is the lack of political support in selling the industry to the general public, which is important in a complex business requiring huge investments and long lead times. One accusation is that politicians are actually lagging behind their constituents' views in supporting the industry. Their 'perception of perceptions' could be faulty and they are frightened to support the industry, even more so than the general public. But politicians should lead the people in such complex issues as nuclear. If the UK needs a dozen new nuclear power plants, the Prime Minister should be capable of standing up in the House of Commons and persuading the public that this is right. But he is unlikely to do this, preferring (as do politicians around the world) limp-wristed statements of retaining nuclear as an option for the future. This essentially avoids considering the true issues and puts them into somebody else's in-tray to be tackled in a few years time. The reason politicians do this is easy to explain.

Most politicians these days are essentially reactive rather than leaders. They use focus groups to test the impact of any policy changes and nuclear will always come up as a big 'no-no' in these. This is because it is very strongly opposed by a small number of people, maybe 5% or 10% of the population. Politicians avoid these types of issues as they know if they strongly push them, the 5-10% of antis will not vote for them, irrespective of any party affiliation. So they are immediately losing a section of the electorate which could be important in a tight poll, possibly costing them victory because of only one issue. Far better to perform a neat feint and talk about something else. An analogous example was the proposal to abolish fox hunting with dogs in England. The Prime Minister, Tony Blair, was personally in favour of abolition, but there was a small minority (again perhaps 5-10% of the population) strongly against. These would almost certainly vote against anyone supporting abolition at the next election. Although hunting was eventually effectively banned by legislation, with strong popular support, Blair and other senior political figures essentially avoided the issue, because they didn't want to be marked out of 90 or 95 rather than 100 at the next election.

Another problem the industry faces is down to the large number of issues under which it comes under attack. If we can satisfy the opposition on one issue, they simply move onto another. This has a cumulative wearing down effect.

On each of the major issues, such as plant safety, waste management, radiation exposure, proliferation risks and transport of materials, the industry may win the individual arguments. It can convince people that they can sleep easily in their beds at night, perhaps with a 90-95% margin of certainty on each issue. However, there always remains a lingering 5-10% of uncertainty. Unfortunately, these 5-10% doubts tend to be additive in people's minds, possibly pushing many people to believe that nuclear power has so many little things running against it that it should be avoided. If there are so many things which can go wrong, surely it is too risky and we should look for an alternative?

Finally, what has been the industry's prime response to its image problem? The solution today is an open dialogue with all the stakeholders, with frank exchanges of views. This is a huge improvement on the former hectoring tone of the industry, essentially with scientists and engineers lecturing the public that they were too stupid to understand the complexities of the arguments and should therefore shut up.

Today's approach is to give people as much accurate information as possible about the industry, on the basis that if they fully understand the facts, they will be more supportive. But the facts are unfortunately insufficient as anti-nuclear people are far from ill-informed. On the contrary, many are knowledgeable but just interpret things differently, with alternative value systems. So who is providing the information is crucial, as is also the style in which it is delivered. Excellent websites, printed publicity, media advertisements and speeches are fine, but the industry vitally needs more inspirational politicians, leading companies with strong public images and third party advocates to speak up on its behalf. This will help dispel the cumulative wearing down effect which has been so pervasive. But above all else, the industry needs to get beyond its smooth talk of a 'nuclear renaissance' and prove it is alive and kicking with a major plant construction programme, based on sound economics. This would do wonders for public perception and we may find that the problem then simply drifts away.

Nuclear's low external costs: justifying public support? Critics of the nuclear industry claim that it has always been heavily dependent on government subsidies and cannot survive in liberalised electricity markets without the beneficial hand of the state. The industry certainly accepts that it has received substantial public funding for research and development (R&D), but most of this was in the past and justified by society's need to develop alternative energy resources and technologies. It is also quite difficult to separate out military-related R&D in the nuclear sector from that aimed at the civil nuclear fuel cycle. Just think of enrichment technology, for example. The civil industry today is clearly benefiting from a great deal of previous R&D expenditure that was clearly, initially at least, intended for military applications.

In more recent years, R&D expenditure from public funds has spread increasingly to the renewables sector, which has also received substantial funding from private industry. Whether all this investment in the future will eventually be rewarded remains an open question – only time will tell. By its very nature, such funding often achieves little obvious direct benefit, but is something that society clearly ought to do if it has a keen interest in the future. Even a few seemingly small breakthroughs may lead to great things.

Renewables also receive additional direct and substantial subsidies from the state. These are justified partly on the grounds that they need assistance in the early stages of development but mainly because they avoid greenhouse gas emissions. The subsidies take various forms, but have been large enough in countries such as Germany and Denmark to make wind power generation an important element in power generation. In the United States, wind power receives a production tax credit of 1.5 cents per kWh, which is also inflation-linked. In the UK, the combination of the Renewables Obligation and the Climate Change Levy has been worth around 5 pence per kWh to wind power stations, which is more than double the average power production cost across all generation technologies.

The nuclear industry argues that it is also very environmentally friendly, as it has always had to incorporate its own waste management and disposal costs in prices charged (equivalent to about 5% of generation costs, with a further similar sum for decommissioning) and doesn't emit greenhouse gases. Fossil fuel producers, on the other hand, receive implicit subsidies as the waste products of energy use are dumped into the biosphere. If nuclear has such small external costs, surely it should receive similar benefits as the renewable sources of energy? On the other hand, the industry's critics doubt that it is so beneficial, mentioning radiation exposures and hidden costs that are avoided.

Questions such as this can be addressed by lifecycle analysis (LCA). An important component of LCA is the study and possible quantification of external costs. The report of ExternE, a major European study of the external costs of various fuel cycles, focusing on coal and nuclear, was released in 2001 and further figures have emerged since. The European Commission (EC) launched the project in 1991 and it was the first research project of its kind "to put plausible financial figures against damage resulting from different forms of electricity production for the entire EU."

The external costs are defined as those actually incurred in relation to health and the environment and quantifiable but not built into the cost of the electricity to the consumer and therefore which are borne by society at large. They include particularly the effects of air pollution on human health, crop yields and buildings, as well as occupational disease and accidents. The 2001 data excluded effects on ecosystems and the impact of global warming, but these are now included despite the high range of uncertainty in adequately quantifying and evaluating them economically.

The methodology measures emissions, their dispersion pathways and ultimate impact. Exposure-response models lead to evaluating the physical impacts in monetary terms. With nuclear energy the (low) risk of accidents is factored in along with high estimates of radiological impacts from mine tailings and carbon-14 emissions from reprocessing (waste management and decommissioning being already within the cost to the consumer).

The report shows that in clear cash terms nuclear energy incurs about one tenth of the costs of coal. Nuclear energy averages under 0.4 euro cents per kWh (0.2-0.7 cent range), less than hydro; coal is over 4.0 cents (2-10 cent averages in different countries); the range for gas is 1-4 cents; and only wind shows up better than nuclear, at 0.05-0.25 cents per kWh average. The European Union (EU) cost of electricity generation without these external costs averages about 4 cents per kWh. If these external costs were in fact included, the EU price of electricity from coal would double and that from gas would increase by around 30%.

The report proposes two ways of incorporating external costs: taxing the costs or subsidising alternatives. Due to the difficulty of taxing in an EU context, subsidy is favoured. EC guidelines published in February 2001 encourage members states to subsidise "new plants producing renewable energy ... on the basis of external costs avoided," up to 5 cents per kWh. However, this provision does not extend to nuclear power, despite the comparable external costs avoided. EU member countries have pledged to have renewables (including hydro) provide 12% of total energy and 22% of electricity by 2010, a target which appears unlikely to be met. The case for extending the subsidy to nuclear energy is obvious, particularly if climate change is to be taken seriously.

Consideration of external costs leads to the conclusion that the public health benefits associated with reducing greenhouse gas emissions from fossil fuel burning could be the strongest reason for pursuing them. Thousands of deaths could be avoided in urban areas each year by reducing fossil fuel combustion in line with greenhouse gas abatement targets. The World Health Organization (WHO) in 1997 presented two estimates, of 2.7 or 3 million deaths occurring each year as a result of air pollution. In the latter estimate, 2.8 million deaths were due to indoor exposures and 200,000 to outdoor exposures. The lower estimate comprised 1.85 million deaths from rural indoor pollution, 363,000 from urban indoor pollution and 511,000 from urban ambient pollution. The WHO report points out that these totals are about 6% of all deaths but the uncertainty of the estimates means that the range should be taken as 1.4 to 6 million deaths annually attributable to air pollution.

An interesting area is that of accidents related to energy production. This should be an area that is particularly obvious to the general public, allowing them to make accurate projections of risks and rewards. Unfortunately it is not so simple.

A November 1998 study from the Paul Scherrer Institut in Switzerland drew on data from 4290 energy-related accidents, 1943 of them classified as severe, and compares different energy sources. It considered over 15,000 fatalities related to oil, over 8,000 related to coal and 5,000 from hydro. Considering only deaths and comparing them per terawatt year, coal has 342, hydro 883, gas 85 and nuclear power only 8. Given that nuclear power annually delivers some 2500 TWh per year to the world, these 8 deaths would be spread over 3.5 years, so on average 2.3 deaths a year. Coal produces slightly more than double nuclear's share of world electricity and the average works out at 216 deaths per year. In terms of number of immediate deaths per event from 1969 to 1996, hydro stands out with about 550 compared with coal at around 40. Nuclear power has, of course, only experienced one accident of such magnitude, namely Chernobyl, despite the common perception that it is so dangerous.

One only has to look at the coverage received by the incidents in the Japanese nuclear industry in recent years to appreciate the problem the industry faces. A small number of deaths in the nuclear industry receive huge headlines even if they are not attributable to the nuclear part of the plant – as was the case with the August 2004 accident at Mihama 3 in which a steam pipe broke, killing five workers. In contrast, frequent coal mining disasters in China, sometimes with hundreds of deaths, receive little attention. Yet both are related to similar kilowatt hours of power received by the consumer.

The recent studies, nevertheless, show that nuclear power is certainly not a bad citizen and in fact, to the contrary, is very safe to work in and has very limited impacts on the external environment. The battle to achieve acceptance of these important facts is a very slow and hard one, but it is one that the industry appears slowly to be winning. The task then is to persuade society to be more rational in its decision-making after people are given the correct information and necessary price signals. The external costs of power production should certainly be incorporated in electricity pricing, but it is maybe over-optimistic for this to happen in such a way as to give nuclear a substantial economic advantage – at least for some time yet.

One alternative is for new nuclear plants to receive some direct assistance from the state, such as production tax credits, loan guarantees and speedier regulation. Anything which gets new nuclear plants over the hurdle of the heavy initial capital investment cost will clearly be helpful and could be justified by helping the environment and maybe also securing a greater balance in power supply.

Nevertheless, the first stage is to achieve the widespread recognition that nuclear is benign. This in itself would go a long way towards countering the problem that nuclear still has with general public acceptance.

Potential weapons proliferation: a real risk?

Those opposed to nuclear power, to satisfy rising electricity demand, national security of supply objectives and the global need to curb greenhouse gas emissions, employ a number of arguments to make their case. The majority of these can easily be shown to rest on faulty understanding of the real situation or just plain bad analysis – for example the claim that the uranium will soon run out or, even more ridiculously, that the move to 'lower grade' mines, requiring heavy energy inputs to process the ore, may actually result in nuclear power consuming more energy than it produces.

There are, however, more respectable arguments the antis can employ against nuclear. Waste management is clearly one of these – it is fine for the industry to say that it has the technical solutions but just lacks the political will to employ them, but this isn't really good enough. We need something better than this. There is also another important area that has rather crept up on the rails as an issue the industry must confront, particularly before it builds lots of new reactors. This is the risk of proliferation of nuclear weapons, with a source in the civil nuclear fuel cycle.

A few years ago, this didn't seem to be such a major concern. The Cold War was over, leaving just one superpower and a small number of other countries with nuclear weapons. The 1970 Treaty on the Non-Proliferation of Nuclear Weapons (NPT) could be regarded as a significant success, indeed one of the few international treaties that had achieved much of what it set out to do. India, Pakistan and Israel (assumed but not admitted) had acquired nuclear weapons outside the NPT but this was not a bad result, given the general fears about nuclear proliferation expressed in the 1950s and 1960s. Many countries had built civil nuclear power reactors without any thought of ever getting involved in nuclear weapons. The nuclear industry could claim that the civil and military sides of nuclear could, indeed, be separated and the proliferation risk, although always present, could be put to the back of most people's minds.

Things have now unfortunately changed. The upsurge in global terrorism, combined with the fears over developments in Iraq, Iran, North Korea and Libya, have refocused people's minds on the risk of mass proliferation of nuclear weapons sourced from civil nuclear power or, at the very least, expropriation of civil nuclear material by terrorist organisations to make a 'dirty bomb', which would cause a significant amount of disruption and possibly some loss of life.

The industry may continue to emphasise the strength of the safeguards system and the effective separation of the civil and military aspects of nuclear, but it is clear that new nuclear build must satisfy the public's fears. This is not helped by the clear hypocrisy shown by some nations. International geopolitics has much to do with this, but it is difficult to get this over to the 'man on the street'.

There are, however, some very positive developments in this difficult area. The new challenges and the variability of political will when confronted with situations such as Iran suggest that moving to some kinds of intrinsic proliferation resistance in the fuel cycle is timely. There are a number of ideas, previously floated many years ago but then seen as too difficult and not really necessary, which have been dug out and revamped. One key principle is that the assurance of non-proliferation must be linked with assurance of supply and services within the nuclear fuel cycle to any country embracing nuclear power. This raises the question of whether multilateral initiatives should be under International Atomic Energy Agency (IAEA) control or coordination so that the IAEA might guarantee the supply of nuclear fuel and services for *bona fide* uses, thereby removing the incentive for countries to develop indigenous fuel cycle capabilities.

Impetus has been given to this by the leadership of Mohammed ElBaradei, Director General of the IAEA, who has pointed to the need for better control of both uranium enrichment and plutonium separation. "We should be clear," he said, "that there is no incompatibility between tightening controls over the nuclear fuel cycle and expanding the use of peaceful nuclear technology. In fact, by reducing the risks of proliferation, we could pave the way for more widespread use of peaceful nuclear applications." This echoes the rationale of the NPT itself, and he has brought these matters to the attention of the UN General Assembly. As well as constraining the 'do-it-yourself' inclinations of individual countries, "multilateral approaches could offer additional advantages in terms of safety, security and economics," he said.

There are several approaches under discussion by an expert group convened by the IAEA, including: developing and implementing international supply guarantees with IAEA participation, for example with the IAEA as administrator of a fuel bank; promoting voluntary conversion of existing facilities into multinational control, including the non-signatory countries to the NPT (such as India and Pakistan); and creating new multinational, possibly regional, fuel cycle facilities for enrichment, reprocessing and used fuel management, based on joint ownership. A further idea is to reinforce existing commercial market mechanisms of long-term fuel supply contracts, possibly involving fuel leasing and the take-back of used fuel, so obviating the need for fuel cycle facilities in most countries.

Such arrangements are very much 'for now' but there are realistic hopes that the 'problem' as it's identified today may eventually largely go away. Generation IV reactor systems with full actinide recycling as part of a closed fuel cycle will produce very small volumes of fission product wastes without the long-life characteristics of today's used fuel, and will have high proliferation resistance. The 'classic' closed fuel cycle with aqueous (PUREX) reprocessing and recycling of plutonium into mixed oxide (MOX) fuel is not intrinsically proliferation resistant; yet there are clearly substantial difficulties in diverting the materials for illicit uses – IAEA safeguards have indeed successfully prevented any diversion and commercial (reactor-grade) plutonium is thankfully most unattractive for weapons.

One barrier to the creation of multinational fuel cycle facilities, with attendant guarantees of supply in exchange for strict adherence to safeguards, is the view held my some countries in the world that they ought to develop full fuel cycle facilities because of security of supply or import-saving reasons. Transport of nuclear fuels has become a difficult issue, to add to concerns about the reliability of various suppliers, while countries possessing significant uranium resources are inclined to develop them and then think about developing other areas of the fuel cycle too. Hence Brazil's involvement in uranium and enrichment, to fuel its own reactors and, less obviously, the regular views expressed in Australia that it should 'add value' to its uranium sales by converting and enriching too.

Apart from the non-proliferation advantages of multinational fuel banks and fuel cycle facilities, it may also be argued that the economies of scale in uranium mines, enrichment and reprocessing plants and eventually waste repositories, suggest that there should be only a small number of facilities worldwide. Although developing small uranium mines and enrichment plants may appear to meet some immediate national objectives, in the long run it would be better from the economic standpoint to re-deploy the resources elsewhere and buy, with guarantees, from abroad. Getting this point agreed by some countries may prove rather difficult but is perhaps best made with repositories (less an issue for proliferation than enrichment and reprocessing facilities). The current national solutions make little sense either economically or politically, yet moving to an international regime requires substantial changes to the rules of nuclear commerce.

It is clear that re-jigging these rules about the movement of nuclear materials, mostly drafted many years ago to cope with markedly different circumstances, is now long overdue and very worthwhile. One stimulus may be the new arrangements on nuclear trade the United States is now seeking with India, so many years out in the cold owing to its weapons programme. This doesn't mean throwing away all the arrangements built up by the Nuclear Suppliers Group (NSG) but they must now be made more flexible to cope with a changed world. The United States has undoubtedly jumped the gun a little and put other nations quickly on the spot but the overall impact is very favourable. Simply categorising the second most populous nation in the world as a nuclear outlaw was never helpful and, if anything, hardened attitudes.

One important matter is now to ensure full and effective verification of the NPT safeguards regime, through universal implementation of the Additional Protocol to each country's safeguards agreement with the IAEA. This gives the IAEA broader rights of inspection and is now firmly established as the contemporary standard for NPT safeguards. In those instances where a confidence and credibility deficit has arisen, additional *ad hoc* measures may also be required.

In conclusion, it can be seen that a lot is happening at present to lessen the general public's reasonable concerns about the proliferation risk from civil nuclear power. Similar to the safety issue, the best advertisement is many years of sound operation of nuclear power plants without anything going wrong. The superb safety record of nuclear reactors and the attendant fuel cycle is now very well-documented and the concerns of the public have definitely been reduced. On the proliferation side, a lot remains to be done, but the correct seeds have been sown. Possible diversion of fissile materials to illicit uses is likely to come up as an issue when any new nuclear build programme is proposed, but it should be possible to satisfy the concerns of those interested.

Nuclear and renewables: can they be partners?

There is a popular view that nuclear power and renewable forms of energy, such as wind, solar and tidal, are competing for a place in the energy mix. Advocates of both are fond of issuing 'knocking copy' about the other, stressing the various difficulties and limitations in them playing an enhanced role in the energy future, while views are getting polarised by carbon emissions becoming a very live popular issue. Given that both nuclear and renewables emit zero or very little carbon, they can potentially play important parts in curbing emissions if they can replace fossil fuels in future generation strategies.

Those pushing renewable energy are insistent that nuclear and renewables are naturally competitive, and cannot coexist as happy bedfellows – therefore nuclear should not have a major place in any future generation mix. They trot out all the usual arguments against nuclear, but it is clear that the real opposition lies somewhere much deeper. The sceptics know that the industry can put up a stout defence on each of these issues – for example reactor safety, the risks of nuclear proliferation and waste disposal. Yet the nuclear opponents will continue to present them as a cover for a deeper complaint. This is that they are essentially opposed to the way the modern world works and wish to move to a much more decentralised economy where energy is supplied on a more localised basis. Nuclear power embodies everything they dislike about today's world – as a large-scale centralised form of generation, with the involvement of big government and large corporations, inevitably somewhat remote from local decision-making. They are deeply uncomfortable with the way the modern capitalist system works, with the trend towards globalisation of production and (implied at least) imposition of outside cultures on local people.

Their alternative would be to move towards a more localised and small-scale form of electricity provision, often called distributed generation. This could ultimately result in people generating power in their own homes, using low carbon-emitting technologies, with the addition of a very localised grid system to take up any surpluses or cover deficits. In some ways, this can be characterised as a rather naïve and romantic vision of a return to the world as it was before we developed centralised power generation, but it is certainly realistic as a very long-term vision of where we may head. Unless very small nuclear units become a workable proposition, nuclear power as it stands today doesn't fit at all well into this vision, hence the strong opposition to expanding nuclear today.

Although it is realistic to propose that rich industrialised societies can try to move away from centralised generation, this cannot happen overnight. They have developed a very different pattern of providing electricity and this will dominate for many years in the future. The most important question is how to provide large quantities of electricity at a few locations, in ways that are economic, environmentally friendly and contribute to security of supply. Renewable forms of energy can certainly contribute to this, but too much cannot be expected of them as they will be limited by both economics and their intermittent character.

For rapidly growing developing countries such as China and India, it is also unrealistic to believe that they can achieve their economic objectives without large-scale centralised power generation, hence they have to seriously consider nuclear if they are to achieve those objectives in a way which minimises carbon emissions. They already have large cities with huge power demands from residential, commercial and industrial demand sectors and further widespread urbanisation is unavoidable. In their rural areas, however, distributed generation is a workable proposition, hence their interest in renewable energy systems as well as large-scale generation. This also goes for many other developing countries, not so far extensively urbanised, where millions of people still do not yet have regular access to electricity. The amount of research and development work directed at renewable energy forms, currently underway around the world, will surely produce a lot of new options for these countries. It is likely, however, that they will still require some generating units of 100-300 MWe and this is where smaller nuclear reactors such as the pebble bed modular reactor (PBMR) could find a ready market.

For major industrialised countries such as the UK and the US, there is also now a significant requirement for replacement capacity, as existing generation facilities are shut down for economic or environmental reasons. In the case of the UK, some of these will be nuclear, as the Magnox and AGR reactors are shut down. Renewables cannot yet hope to replace these shutdown facilities and must await a more wholesale change in the electricity supply infrastructure to gain a greatly enhanced share of the total. It would certainly be useful to experiment with alternative power supplies for newly developed residential areas, so a better idea of costs and benefits can be obtained.

There is also an important claim of the renewables advocates that investing in more nuclear, even if it contributes to curbing carbon emissions, will somehow 'crowd out' investment in their favoured technologies. The past may supply some grounds for harbouring such worries, as the existing power supply systems grew up by strongly favouring one generating mode after another, based initially on coal, then followed by strong successive pushes towards oil, nuclear and finally gas. But there is today a more mature recognition that each power generating technology has significant costs and benefits. It is therefore highly unlikely that any mode should completely dominate production and that a mix of new capacity is likely to be the optimum in nearly all circumstances. In addition, if curbing carbon emissions is the prime objective, we will need to seriously consider every generation technology which could assist this.

Nuclear is well-suited to covering baseload electricity requirements while renewables can add substantial power increments on top. Yet nuclear is somewhat inflexible and doesn't work well economically while load following, so cannot easily act as backup to intermittent generation modes. Renewables themselves contain their own individual mixes of advantages and disadvantages, but some of their proponents are unfortunately guilty of the same arrogant and blinkered thinking displayed by many nuclear advocates in the 1950s and 1960s. For both OECD nations and the developing world, a mixture of generating options is surely appropriate, determined by resource endowments, geography, energy security, environmental and other considerations. To rule out any option through ideology is not appropriate.

It should also be said that determining the generation mix should, as far as it's possible, be informed by reference to the facts. Yet it can't be settled solely with regard to the facts as people have very different value systems and will interpret the available information differently. But the more ridiculous arguments against nuclear, such as that it provides no net incremental energy addition, must be quickly dismissed from the equation. Considering France, with an 80% nuclear share in its electricity generation mix, should be sufficient to demonstrate that nuclear can provide large quantities of cheap and low carbon emissions power. It is abundantly clear that the public still needs better information on energy matters as years of cheap fossil fuels have induced complacency. Yet there is no need for advocates of any energy solution to engage in further knocking copy against the others. The expected growth in world electricity demand, doubling by 2030 according to the International Energy Agency, should leave plenty of room for substantial growth for all modes of generation.

It must also be recognised that emerging energy technologies should receive public subsidies in order to allow them to develop. Nuclear received substantial public backing in the past and renewables deserve the same today. The first new nuclear units to be built should not now need financial subsidies, as the economics now look sound, assuming that investors can take a long-term view. The key requirement is for the public authorities to develop a clear regulatory environment and develop national policies on waste management and decommissioning – new plants can then set aside appropriate funds to cover these future liabilities. Nuclear power does not necessarily have to be, in the future, a creature of 'big government' and need not crowd out a desirable rapid expansion of renewables. Providing cheap and largely carbon-free baseload power 24 hours per day is a sound basis for any electricity system, with other generation options supplying the balance. To some extent, there may be a competition for limited government attention and funding, but if low carbon energy solutions are important, it should not be impossible for government to encourage nuclear and renewables to expand simultaneously.

It therefore makes good sense for the advocates of both nuclear and renewables to try to throw off the baggage of the past and move forward together. Neither is going to go away and so friendly coexistence would seem to be a good policy. If a reduction in carbon emissions is needed, nuclear technology is available today. That it will take some time to build new nuclear plants argues for streamlining the regulatory regime as far as is compatible with meeting reasonable requirements for public review. The argument that new nuclear cannot quickly help curb carbon emissions is somewhat disingenuous when coming from those doing all they can to slow down approvals for new plants. The share of renewables in world electricity is also clearly capable of rising substantially, but it will similarly take time for this to occur. The long-term energy future is very uncertain – if we move to systems based largely on hydrogen rather than hydrocarbons, there are good possibilities for both nuclear and renewable technologies. Yet decisions should be made after careful presentation and discussion of all the facts, and no side has any true interest in these being hidden or obscured.

Public opinion: why hasn't it yet been turned around?

Opposition to anything to do with nuclear is central to the Green creed – indeed, it is an issue such groups have used as a unifying force amongst their often disparate memberships. The roots of anti-nuclear sentiment clearly stem from the links between civil nuclear power and nuclear weapons. The devastating effects of the atom bombs dropped on Japan in 1945 and the fears induced by the subsequent nuclear arms race between the superpowers is something that the civil nuclear sector has never really succeeded in casting off. The campaigns against nuclear weapons in the 1950s and 1960s involved an entire generation of young people from many political persuasions, who found a common cause around which to rally.

The idealists who hoped that nuclear weapons could be removed from the face of the Earth have been sadly disappointed – perhaps not surprisingly, as it's difficult to un-invent a proven technology. But it is arguable that the movement has actually been rather successful on the basis that there has been no subsequent use of nuclear weapons after Japan in 1945, the number of countries possessing them has hardly increased and the quantity of warheads held by the major powers has been reduced by arms limitation treaties. Testing of weapons is now also greatly constrained by treaty.

The civil nuclear power industry cannot escape its obvious origins within the military programmes and there are links still evident today – from the civil side, the most beneficial of these is the downblending of Russian highly enriched uranium (HEU) from former warheads, which is supplying reactor fuel to satisfy around 10% of current US electricity requirements. Many of those formerly marching against the bomb still have deeply-held convictions against any use of nuclear technology; indeed in extreme cases, even the beneficial applications in medicine and agriculture. Nuclear power stations provide a very obvious symbol of something people are, at best, very suspicious of and, in other cases, strongly opposed to. Major incidents, such as Three Mile Island and Chernobyl, are seized upon by opponents as evidence of society's foolishness in playing with the nuclear devil. Minor incidents are publicised as posing significant threats to human health, without any perspective drawn from inferior safety records at other energy businesses.

Supporters of the civil nuclear industry argue that this is all rather unfair. They point to the barriers between the military and civil sides of nuclear and claim that the connections made by opponents are illogical. For example, not many people object to the widespread expansion of civil aviation on grounds that planes can also be designed as formidable fighting machines, able to deliver death and destruction. By any scientific evaluation, the industry's safety record is excellent and studies show that the external costs of nuclear are minor when compared with other electricity generating technologies. The thousands of deaths in coal mining each year, explosions at gas terminals and devastating floods when hydro dams are breached, receive a fraction of the publicity accorded to even minor nuclear incidents. Those in the industry know that journalists like an easy story and nuclear provides this only too readily, as it is impossible to completely avoid every minor incident.

Each of the main arguments used against nuclear, such as safety, waste management, risks of proliferation and economics have been rebutted as far as possible, yet the general anti-nuclear sentiment has been very hard to shift. People who live near nuclear power stations are usually highly supportive, on the basis that the stations provide stable well-paid employment and few problems, but there remains a widespread view elsewhere that nuclear is a risky option and its proponents merely acting out of self-interest.

If the industry's case is so strong, why has it not been more successful at rebutting its opponents? There are four main reasons: historically poor communications; the sheer number (if not the quality) of arguments against the nuclear industry; the deep emotional currents that often swamp consideration of the facts in people's minds; and finally the changes in the political process in key countries.

The civil industry's early communications with its stakeholders (to use modern parlance) were undoubtedly poor – indeed this remained the case until comparatively recently. Arrogant scientists and engineers would address audiences and the media as if they were children – basically saying: "We've developed this marvellous new technology for you, so you'd better go out and use it. Just do as I say!" Memories of "too cheap to meter" are frequently brought up and maybe exaggerate what was generally said, but the obvious potential costs of nuclear were certainly dismissed and only the benefits given any credence. The other arrogance was to suggest that nuclear could eventually dominate the energy world, on the basis that fossil fuel supplies would soon run out and become uncompetitive. It has taken a long time for the industry to live all this down, with attempts to use modern communication techniques taking time to bear fruit. This is now gradually happening, but it is proving a long haul to overcome the legacy of the past. It is not a matter of slick industry salesmen in sharp suits now replacing the well-meaning but incompetent scientists of the past, but more of having eager willingness to engage with all groups of society and patiently explain both pros and cons of nuclear and other technologies.

The industry argues that each of the key arguments used against nuclear technology have very little merit. In each case it may well be realistic to persuade 95% of the people of this. Or alternatively to persuade everybody with a 95% degree of certainty in his or her mind. Yet the residual 5%s remain very important, because they are additive. The 5% doubters in the population on one aspect (for example, risks of nuclear proliferation) may be an entirely different group from those concerned about another issue (maybe plant economics). So with several separate arguments used against nuclear, it is not difficult for opponents to achieve a large number of doubters in the population – maybe not 50% but at least a significant minority. Alternatively, the 5% elements of doubt in any individual's mind on each issue are similarly cumulative. Lots of 5%s begin to add up to the extent that many people will say: "Well, there has got to be something wrong with this technology, as so many little things can go wrong – so let's use something rather simpler." This wearing-down process accounts for much of the anti-nuclear movement's success – no matter how many arguments are rebutted, there always seems to be another one.

The industry has put a lot of effort into presenting the facts about nuclear power and other power generation technologies, through establishing good websites, providing media interviews and addressing conferences and other interested audiences. This has certainly helped counter some of the more unreasonable claims of the anti-nuclear movement, but it has not been enough. More third party advocates are needed but these have only comparatively recently emerged, notably some formerly identified as leading environmentalists. The bigger problem, however, is that the debate cannot be answered only by reference to the facts. It is conceivable that both sides can agree the key facts, but the interpretation of these and their meaning can differ appreciably. This is because of different views on risk-taking and the values one ascribes to aspects of the world. Careful examination of, and attempting to agree on, the facts will undoubtedly help, but it cannot resolve the matter. For some people, a 1% chance of a nuclear accident in the UK over the next

100 years causing 100 deaths may be completely unacceptable, but can be taken easily in their stride by others who know of the extent of coal mining deaths each year in China.

It is clear that nuclear power needs top-level political support to prosper in any country. Not, note, financial support, but at least the establishment of a reasonable licensing and regulatory regime, defending the interests of all parties, plus clear policies on aspects such as used fuel management and plant decommissioning. Uncertainties on these are fatal to a technology requiring heavy upfront investment followed by many years of operation to recoup these costs before a profit can be made. Yet such support has become hard to win as politicians have generally become more reactive, responding to focus groups and the like, rather than being strong conviction-led leaders. They know that nuclear is an issue that gets a small percentage of the population very excited, either pro or con. So if the government comes out strongly in favour of new reactors, for example, it is likely to lose the votes of all those fiercely opposed to nuclear, irrespective of other considerations in the next election. These votes could be crucial in a tight ballot; so nuclear is, for politicians, a dangerous issue. Thus it tends to be swept under the carpet through fence-sitting, putting off energy reviews until later and so on. We have only recently begun to see the reversal of this, particularly with the Bush Administration's strong support for nuclear in the United States, but it may take greater general public acceptance before other politicians put their necks on the line.

Does this then leave one pessimistic about the nuclear sector's ability ever to win much better general public support? Not necessarily. Indeed, all industries feel that the general public is against them, or at best somewhat sceptical about their activities. Obtaining planning consents for expanding operations always requires a lot of care and attention and providing new jobs in an area is not necessarily sufficient to gain support. Nuclear is not exceptional here and should never expect to be loved. Industry participants often seem to have a chip on their shoulder on this aspect, whereas they should be more confident and ready to move onto the offensive. Producing power for the masses is hardly a glamourous activity but the first point to demonstrate is how important it is, both in the developed world (where electricity demand continues to rise) and in those nations where power supplies are either non-existent or subject to interruption. The difficulties nuclear faces in getting across its case, highlighted above, can best be countered by quietly continuing the years of safe, economic operation and gradually presenting the facts

PUBLIC ACCEPTANCE

in good communication programmes. The complexity of the technology is a barrier, given the low attention span and impatience of both the general public and their elected representatives, but we must continue the effort to explain. The best approach is carefully to explain that each energy option has obvious costs and benefits and nuclear just has its own unique set. Its merits, however, mean that it should be carefully considered amongst every country's energy strategy.

The serious opposition to nuclear: how can it be handled?

People in the nuclear industry are used to the attacks from organisations such as Greenpeace and Friends of the Earth, whose opposition is a deep-seated central tenet of their creed and unlikely to be overturned. The main point is that nuclear embodies everything they hate about modern society and unless we can reshape to fit in with their desire for a small-scale, decentralised energy system, without government and big companies at all involved, the industry has little hope of changing minds. The best it can hope to do is to marginalise them, in the hope that sensible people will realise that their full crazy vision of a 'brave new world' is both unrealistic and, more importantly, actually contrary to what most people actually want.

There is, however, a rising chorus of more substantive attacks on nuclear coming from more intellectually respectable and robust quarters. These have been largely silent for a quarter century, with the contented vision that nuclear will slowly wither away as existing plants gradually shut down and no new ones are contemplated. Now that talk of a 'nuclear renaissance' has spread from the industry to popular debate in the general media as many countries consider new nuclear build (including new countries in all continents), the doubters have reasserted themselves.

Typical of this is the April 2007 report by Charles D. Ferguson for the US Council on Foreign Relations, *Nuclear Energy: Balancing Benefits and Risks*. This is a hugely disappointing document, illconceived from start to finish, more particularly so as senior figures closely associated with the US industry sat on the Advisory Committee. As such, it reads very much like a précis of the book *Insurmountable Risks: The Dangers of Using Nuclear Power to Combat Global Climate Change*, written by Brice Smith for the Institute for Energy and Environmental Research in 2006.

An essential assumption of these publications is that the position of nuclear has improved mainly because it is being advocated as an important answer to global climate change. To some extent, the industry has asked for this, by not spending enough time and effort ensuring that the security of supply and economic arguments are put across as forcibly. The identification of nuclear with clean energy and low emissions is necessary, but leaves open the attack that nuclear cannot be much of an answer as new plants take so long to build that, by the time they are built, a combination of energy conservation and renewable energy may have provided the answer anyway. The Council on Foreign Relations report majors on this, claiming that the rapid nuclear growth scenarios postulated by various bodies are unrealistic. A new argument is that the initial possible shortages of men and materials that new build will encounter will turn out to be a long-run phenomenon and will have adverse economic implications. Costs of new plants will necessarily escalate, pricing nuclear out of the market.

This goes against the previous history of nuclear in the 1970s and 1980s, the experience in many other industries and rational economic analysis. Two hundred reactors came online during the 1980s (39 in 1984 alone), which suggests that a rapid build-up from a low level (and with well-proven technology) is eminently possible. An important point is that despite its technological maturity, nuclear new build will almost resemble a new industry, as there is the need for a huge amount of new training and investment in facilities to manufacture key plant components. The men and machines supporting the previous era of new build are now largely gone so substantial investment is required. This will undoubtedly come once a substantial bank of new plant orders is received, giving long-term security to the investors. With much more reactor standardisation than in the past, modular construction and simpler designs, it is almost certainly that series production will cut and not escalate costs. The first units will experience substantial 'first of a kind' costs, but as the supply infrastructure is rebuilt, nuclear should become more economic and not less. This fits in with the experience of other industrial sectors where mass production of limited lines creates good economics – this can be seen from aircraft manufacturing to the auto industry.

Assuming nuclear generation can expand rapidly (for example to almost 20% of world electricity by 2030 with a doubling of capacity – as in the World Nuclear Association's upper scenario) it can clearly be a substantial contributor to greenhouse gas abatement. The more sensible critics avoid the arguments on uranium availability and carbon emissions from the nuclear fuel cycle, as they know they don't stand up to serious analysis. But it is important for the industry to accept that nuclear can only be a partial solution up to 2030 and not to criticise other programmes which will assist, notably in energy conservation and in promoting sources of renewable energy. But beyond 2030, the gloves can be off, as a combination of Generation IV reactors and a move to a hydrogen-based transportation system could allow nuclear to play a much enhanced role.

Even assuming new nuclear build can be useful in greenhouse gas abatement, the intellectual argument against then centres on battering the reader into submission by listing a long list of possible problems with building a lot of new reactors. This is also a familiar tactic of the wilder Greens – although the events postulated may be low probability, the range of risks is considerable and may be cumulative in the minds of average members of the public. In each case, the worst possible consequence is considered, with little balance drawn from observable evidence in the real world. It is now often grudgingly accepted that reactor safety has been excellent since Chernobyl, so the emphasis has moved onto security of plants, possible terrorist use of civil nuclear materials and alleged weaknesses in the non-proliferation regime. These issues are considered to be additional external costs of nuclear, but sensible energy policy cannot be founded on stoking up fears of low probability but potentially high consequence events. A more rational approach is appropriate, taking into account a full range of advantages and problems.

The other (more traditionally alleged) external costs, such as waste management and plant decommissioning, are also targeted by the critics. The industry claims that these are already internalised by plant operators making appropriate payments and provisions (essentially funded by electricity customers). This is in contrast to the position in fossil fuel power generation, where carbon emissions are not (until now) heavily penalised. It is becoming clear that the announcement of the Global Nuclear Energy Partnership (GNEP) programme in the United States has led to some confusion amongst less-informed observers. It obviously has several possible objectives, notably reduction of new waste volumes and using up existing wastes so that repository needs are minimised, but also strengthening proliferation resistance of the fuel cycle, yet the Council on Foreign Relations report and others don't seem to understand this and get hung up on one aspect or another. In particular, the renewed possibility of fuel reprocessing brings out all the old fears about plutonium but not all countries are likely to go down this road, notably those like Finland and Sweden, which have progressed much further towards operating repositories.

Much of the debate about nuclear at this more sophisticated level comes down to values and interpretations rather than facts. Even when it is possible to agree on the facts, different people have alternative perceptions of risk and it is this that lies at the heart of everything to do with nuclear. It is a complex technology and brings forward a wide range of issues which act like a thick fog in people's minds. Yet the financier of a new plant is in much the same position as someone who lives just down the road from a proposed site for a new reactor or a voter much further away who is presented with nuclear as a serious energy option. The financier has a long list of risks, which must be competently allocated amongst the stakeholders in the plant to give him sufficient comfort to proceed, and without imposing a damaging risk premium on his money. Some are the responsibility of national governments, some will be taken up by the plant vendor and contractors, while others will lodge with the power company itself. The local resident faces different risks, but needs satisfaction on safety, radiation emissions, plant security and eventual decommissioning of the site. The national voter, however, is maybe more concerned by possible proliferation, terrorism and waste management issues.

There are clearly different issues for different groups, but each requires a great deal of industry attention to give them comfort. There is little alternative to increasing knowledge and understanding of the complexities of nuclear, in the hope that the essential audiences will be patient listeners and not feel overwhelmed. It is clear from everyday life that attitudes to risk vary considerably, so even the best industry explanations are unlikely to satisfy everyone. Even confronted with strong facts, some people will always reject nuclear on grounds of major accident scenarios, even if it costs them more for their electricity and means society will need to cut carbon emissions by other means. That is their right, but the hope is that their numerical strength will be low, and that most people can begin to see though that dense fog.

The battle to win hearts: how can it be won?

It is often said that winning over public opinion about its operations is the biggest and most important challenge facing the nuclear industry. When people say this at conferences, industry people tend to sigh and nod their heads, often without really thinking much about the issues. It's kind of taken as a given – the industry generally has a bad public image and this places constraints on its operations, making them more complex and expensive, notably by the imposition of an over-prescriptive regulatory regime.

But are things really as bad as is sometimes made out and is putting a major effort into managing public opinion really so important? The answer is probably not. If we look at the situation in the United States, it is obvious that the vastly improved public perception of nuclear power has its roots in the superb operating performance of the 104 units in recent years. Producing large quantities of electricity cheaply, safely and with regard for the environment is far more effective than any fancy communication strategies. It is only when things start to go wrong at the operational level that the public becomes interested. Hence the incidents at the Brunsbüttel and Krümmel nuclear plants in Germany in 2007 and the earthquake which affected the Kashiwazaki-Kariwa reactors in Japan during the same year show the need for good management of public opinion and demonstrate that the industry still has a lot to learn about crisis management.

In fact, the public doesn't usually have much interest in energy matters as a whole and only tends to get involved when there's a crisis. If the lights go off or there are queues at the petrol station, people get highly upset and put huge pressure on industry and the politicians. But the 1980s and 1990s were a relatively quiet period; so most people today haven't any strong and well-developed opinions about a particular fuel or energy strategy. More recently, however, we have had rapidly-rising fossil fuel prices and concerns about energy security of supply have resurfaced. But it's probably the relationship between energy use and the environment which has begun to touch the general public's consciousness. Climate change and potential global warming has been a gift to the environmental movement, as it presents a more credible apocalypse scenario. Most sensible people recognise that the other fears they have stirred up are groundless as, in general, economic growth can be seen to lead to a cleaner environment, but this is something new and potentially scarier.

Putting nuclear power into this perspective, there are clearly concerns in the public mind about the weapons link, over proliferation coming from the civil side of the industry and a general fear over radiation releases from operations. We can put some of this down to an irrational evaluation of the risks involved but this is something the industry has to live with. The number of people who have a hardened belief against nuclear and will be very difficult to sway are fortunately relatively few. The fact that many people haven't had to think very hard about energy matters for some time suggests that opinion can easily be influenced one way or another. Unfortunately, we can largely forget about politicians demonstrating any degree of leadership in this area. We know from bitter experience that they prefer to sit on the fence when it comes to any matter which can excite even a very small part of their electorate, as losing these committed votes could be crucial in a tight election. So they rely on focus groups and tend to be led by the public rather than vice versa, arguably the opposite of what they're supposed to do. But climate change provides a fantastic opportunity for nuclear to be seen in a new light by those who have some general, but not deep-seated, concerns about it. Presenting it as a green and friendly technology is going to take time, but the message that nuclear emits few greenhouse gases seems to be slowly getting across.

Many of the problems the industry has with public opinion can be blamed on the sins of the past. Indeed, when things go wrong today, as happened in Germany and Japan, it is clear that the lessons from the past have not been adequately learned. Arrogant spokespeople, talking down to their audience and not being open with important information is a legacy the industry has taken a long time to shake off. Unfortunately we can see today that it still hasn't quite got there. Society itself has now changed substantially and nuclear has had to fit in with this. The 1950s to the 1970s were characterised by state provision, deference and a belief that the application of science could bring the greatest good to the greatest number. But from the 1980s onwards, self-reliance, distrust of science and assertion of individual rights irrespective of the common good has become prominent. In itself, nuclear power doesn't sit easily with this, as it relies on a degree of state involvement (at the very least in setting a framework for its operations in licensing, regulation and waste management) but is at last learning to exist within a climate of competitive power markets and private ownership.

The best examples of winning over people in today's world come from specific examples of planning new facilities, rather than attempts at the general persuasion of the wider general public. In Sweden and Finland, the siting of the waste repositories and the fifth Finnish reactor demonstrated that careful work with local people can bring huge dividends. The need for the new facility must first be shown convincingly, and then the public brought into the full process with the provision of clear information and opportunities for consultation. The local people must be respected as the experts in local matters and should ultimately have the final veto on the project. The companies concerned must be seen to be interested in far more than profit and be seen to ultimately have the interests of the local area and the wider country at heart. At the local level, nuclear facilities offer well-paid and secure jobs for many years in the future and have widespread economic impacts beyond the immediate investment.

The industry has identified the provision of clear and accurate information about nuclear power in general and its specific operations in particular as an important weapon in winning the public over. While knowledge is clearly better than ignorance, this approach has severe limitations and cannot be expected to achieve very much, particularly in the shorter term. An obvious observation is that some of the strongest critics of the industry are, in fact, very well-informed. Indeed, easily the best website on uranium mining throughout the world is run by the World Information Service on Energy (WISE), a strongly anti-nuclear organisation. So there must be a lot more to it than the facts. Beliefs and values are arguably even more important than solid information. If you've taken in an argument by emotional appeal (*e.g.* nuclear power is evil), you're unlikely be swayed by facts that counter that belief – indeed, the opposite may in fact be the case. As many people have accepted a message that nuclear is bad, it will take a lot of effort and a long time to overcome it – lots of people have got to be persuaded to change a view that has been entrenched for many years.

The messenger and the way the message is delivered are also very important considerations, hence the search for credible third party advocates. Industries are seen as essentially self-interested by a cynical public – "they would say that, wouldn't they" – but prominent environmentalists such as James Lovelock and Patrick Moore are worth their weight in gold when they speak up in support of nuclear's importance.

But it's still an uphill battle and some people will never be persuaded. Nuclear power embodies all that some groups hate about the modern world – the application of science, big government and large organisations globalising production. Their arrogance mirrors that of the early nuclear pioneers – they feel they are saving the world for the rest of us, who should follow them like sheep.

Finally, it should be noted that the use of language is very important too. Again the industry suffers today for the errors of the past. If you ask anybody which words they associate most with 'nuclear', they will most likely say 'bomb', 'explosion' or 'war' and not 'power'. The association with military uses is very hard to shake off – had nuclear power alternatively (and more correctly) been termed 'fission power', the difficulties over public acceptance would undoubtedly have been rather less. Although it's now rather too late to quietly re-brand the industry, it's a lesson for the future – be careful in what you casually say, as people are receiving messages beyond what you immediately intend. The other good example is the careless terming of everything coming out of the back of a reactor as 'waste'. This then ensured there would be a requirement that something is done about it in this, or certainly the next, generation. As an alternative, referring to 'used fuel' would have highlighted the possible economic value, so the time period could potentially be much expanded (under the guise of passing on an important asset to the next generation rather than a liability). Other nuclear terms such as 'fast breeder' are less than ideal from the public perspective, conjuring up images of sinister *Dr Strangelove* scientists at work, whereas others, such as 'pebble bed' seem more benign. So it's important that scientists don't have a monopoly in the terminology – it's not necessary to bring in highly-remunerated image consultants for the industry, but some thought of the impact on public opinion should ideally be taken.

In conclusion, experience has taught us that there are a number of ways in which we can contribute to the industry obtaining a more favourable public image. But talk of this being the biggest single issue confronting the future is surely nonsense. The most important thing is to carry on operating the existing nuclear power plants as well as possible and the message will eventually get across. There is a variety of other PR initiatives which can be taken to support nuclear, but none of them is as important as better crisis communications when things do go wrong, as inevitably they will from time to time.

4. NUCLEAR FUEL

Introduction The relatively low cost of nuclear fuel and its stability has always constituted a prime economic advantage of nuclear power, at least by comparison with generating plants fired by coal, oil and gas. This does not, however, mean that the nuclear fuel sector is unimportant or lacking in interest. Indeed, the opposite is very much the case, partly because the front end of the fuel cycle is, in itself, quite complex, with individual markets for each of uranium, conversion, enrichment and fuel fabrication.

Nuclear fuel demand is, over the medium and longer term, somewhat predictable and robust compared with the demand for many commodities. Once nuclear reactors are started up, they tend to run at high capacity factors for many years – indeed, nuclear economics are dependent on this happening. There are, however, different definitions of the demand for nuclear fuel that need to be understood. The biennial World Nuclear Association *Market Report* has become the 'bible' of nuclear fuel market analysis in the period of over 30 years over which it has been published. It represents the views of industry participants themselves and presents a coherent view of future demand and supply trends. The 2007 edition shows a demand picture little changed from the previous report two years earlier but the relative likelihood of the various scenarios becoming reality has clearly shifted. The upper scenario, which would have been regarded as dreamland by many observers five years ago, is now very much a mainstream view today. More information is now available on the timings of new uranium mines but secondary supplies should not be forgotten, as they'll remain an important part of the market for many years yet. The assessment of the overall market balance shows that the market should be adequately supplied in the period to 2020 but additional prospective mines may be needed thereafter.

One question frequently brought up, particularly by those opposed to nuclear fuel, is the adequacy of world uranium resources to satisfy growing reactor requirements. This rests on a misunderstanding of the dynamics of exploration for and proving of new resources, which respond to any perceived tightening in the market. Uranium is certainly far from scarce in any geological sense and there are also significant potential nuclear fuel resources in thorium, even if uranium is fully exploited from longer-term and more expensive resources such as phosphates. A far bigger issue is getting the uranium out of the ground and turning it into marketable production.

The significant surge in uranium prices since 2003 has not yet been reflected in major production increases, but these are now on the way. The price upsurge saw production effectively fixed in the short term as the mines that had survived the long period of depressed uranium prices were already operating at high capacity utilisation rates. The world mining industry has also been going through a tremendous boom, fuelled by demand from China in particular, so men and materials have been in short supply. Production is now, however, beginning to rise sharply, led by Kazakhstan and some African mines. Over time, this will likely spread much further as additional deposits are developed.

If the more expansive scenarios for nuclear power are correct, the period beyond 2015 will require a further sharp increase in world uranium production. Secondary supplies, particularly downblended highly enriched uranium (HEU) declared surplus from weapons programmes will be a diminishing component of supply by then, so further mines will need to be developed. The current boom in so-called junior uranium companies may form the foundation for these. Most of the 400 or so companies that have appeared so quickly are engaged solely at the exploration stage and it will take many years to bring new discoveries into production. Much consolidation will take place amidst this market sector, but the activity today will likely result in much enhanced production prospects, particularly in the period beyond 2020.

The enrichment sector is the front end fuel cycle area which is sensitive from a nuclear proliferation standpoint, and has been marked by significant attention from the developments in Iran, North Korea and elsewhere. New international proposals have been introduced to deal with this but enrichment itself is undergoing some important changes. The most obvious is the gradual replacement of the old gas diffusion technology facilities by centrifuges and the more distant possibility of laser enrichment. Demand for enrichment is also increasing significantly with higher uranium prices, as uranium and enrichment services can be partial substitutes. Finally, enrichment has been the subject of significant trade restrictions in recent years, which are at last gradually unravelling.

The fuel fabrication sector is a rather different area of the fuel cycle, offering a very specialised, high technology service to buyers, rather than the provision of a bulk homogeneous commodity like the uranium, conversion and enrichment areas. As such, it has several interesting features, notably a very competitive market based on facilities located closer to end-users and also a significant degree of corporate consolidation in recent years.

Finally, the upsurge in world uranium prices and the subsequent sharp fall back from the peaks have in themselves created a lot of comment and interest (particularly from the financial sector). To some extent, the price boom was predictable, but it had been delayed for so long that people were beginning to believe that it would never happen. Clearly the market works in a very imperfect way and failed to give the signal to stimulate necessary production increases before it was almost too late. The extent of the price upsurge, far greater than anyone anticipated, has now led to much discussion about how the market could work better in future, focusing on the lack of liquidity and transparency that are apparent. It has also stimulated a renewed look at alternatives to the traditional ways of buying and selling nuclear fuel and encouraged buyers to press for the opening up of as many sources of supply as possible.

Nuclear fuel demand:
will it carry on rising?

With all the attention paid to uranium prices and the need for new primary production, the demand position is largely taken for granted, with the expectation that it will remain very robust and continue increasing slowly. The prospects for nuclear power, however, remain subject to great debate, so it is useful to review the demand side of the nuclear fuel equation.

First of all, it is important to note that the demand for nuclear fuel is really four separate demands. Reactor operators require enriched uranium fuel rods, but in order to get these they have traditionally bought natural uranium concentrates, then contracted with service providers to have this converted, enriched then fabricated into fuel. Most of the analysis of nuclear fuel demand, however, centres around discussion of uranium and enrichment, as they can be at least partial substitutes for each other.

It is also necessary to be careful about different definitions of what is meant by the demand for nuclear fuel. The alternative measures relate mainly to the time period that is being reviewed.

The World Nuclear Association (WNA) and other market analysts use a measure of demand called reactor requirements, which is the amount of fissile material and fuel cycle services that will be required to prepare the fuel that will be physically loaded into reactors. A reactor operator may choose to meet its reactor requirements in a number of ways. Most is normally supplied from annual procurements of fresh fuel, which constitute an alternative measure of demand.

Procurements can be defined as the amount of uranium and fuel cycle services, which a reactor operator receives in a given year from any source, except that taken from its own inventories, including recycled material. Another term for this could be deliveries, with most coming from multiannual contracts with suppliers. However, a significant part of the nuclear fuel market consists of other types of transactions, including spot purchases and borrowing of material.

Most reactor operators hold inventories of material in various forms, from natural uranium to fabricated fuel and these may represent several years of forward reactor requirements. This means that deliveries under its procurement contracts in any year may be substantially above or below its reactor requirements, as it builds up and draws down inventory. Procurements are therefore likely to be a more volatile measure of nuclear fuel demand than reactor requirements and should have a closer relationship with annual uranium production. In an era in which inventories have been drawn down, as over the last 20 years in the commercial nuclear fuel market, procurements will be significantly below reactor requirements.

The recycling of spent nuclear fuel provides a slight further complication. Those reactor operators with reprocessing contracts may use recycled uranium or plutonium to meet part of their reactor requirements. The WNA regards this in exactly the same way as inventory draw-down, in other words as an addition to supply rather than a reduction in demand. Some forecasters, however, deduct mixed oxide (MOX) fuel and reprocessed uranium (RepU) from their measure of reactor requirements, providing a figure net of recycling.

Finally, there are some shorter-term measures of demand. These result from the fuel contracts and their flexibilities whereby a reactor operator can choose to vary deliveries from the nominal amount of the contract either upwards or downwards, depending on his and the market's position at that point. In the short term (the next few years), most reactor requirements are covered by contract commitments, but going further ahead there are substantial quantities which are not. Even the upper flexibilities of contracts will not cover all the requirements during this period and this is something which drives the spot nuclear fuel markets, where the balance has to be made up by available supplies. Estimates of so-called 'uncommitted' and 'uncovered' demand are made by nuclear fuel brokers and traders and are a further useful source of information.

Given that the prime interest is in the longer-term development of nuclear fuel demand, how are reactor requirements calculated? The amount of uranium and fuel cycle services required to operate a reactor is not a simple function of the number of reactors in use and their generating capacity. The operating characteristics of reactors are also important, including the load factors achieved, the fuel enrichment and burn-up levels and also the tails assay in the enrichment contracts. Load factors have been increasing steadily, pushing up nuclear fuel demand more quickly than nuclear generating capacity, but the other factors have largely cancelled each other out over time. Load factors will from now on not rise so much, as practical limits have been reached in many countries, so the key to forecasting fuel demand accurately is getting generating capacity forecasts correct.

The biennial WNA *Market Reports* include three demand scenarios, which are built up at a country or regional level and based on consistent underlying base assumptions. In the 2007 edition, the reference scenario is largely a continuation of recent experience, with increasing numbers of new reactors coming into operation in the period to 2030, but no substantial premature closure programmes for existing reactors. World nuclear generating capacity rises by about 1.5% per annum up to 2030, from 372 GWe in 2007 to 529 GWe in 2030. World uranium requirements rise from 64,000 tU in 2007 to 109,000 tU in 2030, a slightly higher rate of growth. The upper scenario switches the openings/closures balance in favour of a substantial number of new reactor starts; closures are limited to those already announced, such as the Magnox reactors in the UK, while the anticipated phase-out programmes in Germany and Sweden don't take place. World nuclear generating capacity is anticipated to be 730 GWe in 2030 and uranium requirements 149,000 tU. The lower scenario goes the opposite way, with new reactor starts limited to those already firmly announced or under construction, while there are many premature closures of reactors, for either political or economic reasons. Nuclear generating capacity falls to 285 GWe by 2030, with uranium requirements down to 52,000 tU.

There is therefore a substantial spread between the upper and lower scenarios, amounting to almost 100,000 tU by 2030. This seems to be an alarming amount, but is largely a consequence of the long time period adopted and very different views of nuclear's future. There is naturally, however, a tendency to focus on the middle or reference scenario, but they are very carefully drawn up on the basis that each is entirely plausible and, as far as possible, equally likely given the frequent volatility of the energy world. There is also potentially more upside in the upper scenario than downside in the lower scenario. The decline in uranium demand in the lower scenario is not dramatic over a 20-year term and justified by the general expectation that most of the 439 reactors currently operating will have long and maybe extended lives.

There is a very interesting issue within nuclear fuel demand concerning the tails assay. Reactor operators require enriched fuel to supply to the fabricators and can choose a balance of uranium and enrichment services to achieve this. The decision on the contractual tails assay is motivated by the relative prices of uranium and enrichment services and for each price relationship, there is an optimum rate. The WNA scenarios were all produced on the assumption that tails assays for Western reactors will remain at 0.22% to 0.25% in the period to 2030, with Russian-origin reactors at 0.10% (owing to the surplus enrichment capacity available there). With uranium prices now highly volatile, there may be opportunities for Western reactor operators to vary the contractual tails assays outside the forecast range, therefore supplying more or less uranium to the enrichment plant and using variable SWUs.

This could raise or reduce uranium reactor requirements for Western reactors by up to 10%, but there is good reason to expect the effect to be rather lower than this in magnitude. It depends on there being sufficient enrichment capacity, which is uncertain in the West (while Russian plants are still subject to trade restrictions). With the heavy investments currently taking place in new centrifuge enrichment plants, the enrichment price is also very hard to assess. Centrifuge enrichment uses far less energy than the old diffusion technology, but investors will require a good financial return on the new plants that are replacing those where the investments were amortised years ago. Enrichment companies had also been accustomed to using a lower operating tails assay than that implied in contracts, a process known as 'underfeeding' which generates additional uranium supply for their own use. Finally, having lower tails assays may limit the economics of sending tails to Russia for re-enrichment, a practice that has been common in recent years, again creating more effective supply.

In conclusion, we may say that by comparison with the demand for most goods and services, nuclear fuel demand is indeed very robust and is likely to carry on rising, for a few years at least. Vendors need not fear wild swings in demand. After then, it all depends on which demand scenario is nearest the mark, but the downside seems quite limited. Within the overall demand picture, there may be some interesting shifts between uranium and enrichment requirements, but this is a complex matter and hard to judge. It may certainly also be quickly reversed if enrichment prices change substantially.

NUCLEAR FUEL

WNA Market Report: calling the shots?

The 2007 World Nuclear Association (WNA) *Market Report*, titled *The Global Nuclear Fuel Market – Supply and Demand 2007-2030*, continues the long tradition of biennial reports from the WNA and its predecessor organisation, the Uranium Institute, from the time of the latter's foundation in 1975. As such, the report very much represents the views of industry participants on the likely future of the nuclear fuel market, but without being perceived as a lobbying document on behalf of the industry. It points out the challenges as well as the opportunities for nuclear power over the next 25 years, as well as the undoubted opportunities and specifically those which are relevant to the nuclear fuel business. Over its 13 editions, the report has become the main reference point for those interested in its subject area, not just those within the industry but also journalists, financial analysts and business planners. It is available cheaply to non-members of the WNA (certainly compared with consultants' reports), so the arguments and scenarios developed are much-quoted elsewhere.

The 2007 report came at a very interesting time for nuclear power and more particularly for the world uranium market. Nuclear is now very much back on the agenda with talk of a 'renaissance' – the foundation being the much improved operating (and financial) performance of the existing 439 reactors around the world but with the expectation now for many new reactor orders, both in countries already having nuclear power but also in some new countries too. For a variety of reasons, the spot uranium price has soared, peaking at $138 per pound in June 2007, after spending the entire 20 years prior to 2003 at around $10 per pound. It has since fallen back (to around $70 per pound in April 2008) and is increasingly giving the impression of an unpredictable roller coaster ride.

After an introductory chapter describing the main features of nuclear power today and of the nuclear fuel market (aimed largely at the non-specialist audience), the report adopts the traditional route of examining the demand and supply sides of the market separately (Chapters 2-4) before bringing both sides together in Chapter 5, to form some conclusions.

The nuclear generating capacity scenarios in Chapter 2 may surprise some readers as they are little changed from the previous report in 2005. Countries which are now higher (particularly Russia where the recently announced plans provide more grounds for optimism) are generally balanced by others which are lower (such as Germany, where the election of the avowedly pro-nuclear Christian Democrats has not, so far at least, allowed a revocation of the nuclear phase-out, which could take Germany out of nuclear entirely by 2022). Indeed, the greater optimism surrounding nuclear in general was arguably completely incorporated in the previous report – all that has changed is the likelihood of the various scenarios becoming reality. The upper scenario, where world capacity doubles from 368 GWe in 2006 to 730 GWe in 2030, is now becoming increasingly likely, whereas the lower scenario, where capacity is almost flat until 2020 then falls away to 285 GWe by 2030, is logically rather less so. The report stresses, however, that each of the scenarios is internally consistent across the various countries and is worthy of readers' attention – anything regarded as totally unrealistic would be excluded. Given that there are three scenarios, it is perhaps natural that most attention tends to focus on the reference scenario, where nuclear generating capacity reaches 529 GWe by 2030, a rate of growth of 1.5% per annum. Yet even this requires a break with the recent past – as well as operating life extensions for nearly all existing reactors, it requires new reactor startups in many countries, in particular beyond 2020, but in the United Kingdom and United States even before then.

The nuclear generating capacity scenarios become uranium, conversion and enrichment requirements scenarios in Chapter 3, through incorporating parameters on how reactors are operating (such as capacity factors, fuel enrichment levels, fuel burn-ups and tails assays at the enrichment plant) in a MS-Excel based model. Much of the information here comes from a questionnaire sent out to fuel cycle participants, but supplemented by other data and the judgement of WNA members. As uranium has become relatively more expensive in recent times, it is clear that reactor operators have had a substantial incentive to economise on its use. Demand for uranium is clearly inelastic with respect to price, at least in the shorter term, but through lowering the tails assay (thus cutting uranium demand but boosting enrichment demand) and by changing reactor operating cycles (allowing lower enrichment levels) the operators can potentially save a lot of money. In the WNA's evaluation, the result is that the demand for uranium will be essentially flat in the period to 2010 at around 65,000 tU per annum and only begin to pick up in the period thereafter in the reference and upper scenarios. However, by 2030, the upper scenario is at 149,000 tU and the reference scenario 109,000 tU, representing faster rates of growth than nuclear generating capacity itself. This is explained by the expectation that countries currently lagging on capacity factors will eventually improve substantially, while the general trend in enrichment levels continues to be upwards.

The uranium reactor requirements scenarios are overall rather lower than in the 2005 report, while the enrichment requirements tend to be higher. This is essentially because of the tails assay assumption made in the new report – as opposed to the 2005 report's universal 0.27% for Western reactors to 2030, it is now anticipated to fall to 0.22% by 2010 and then rise again to a constant 0.25% in the period 2015-2030. This is a very important assumption but one that is strongly supported by the questionnaire responses and by other market intelligence.

Turning to the supply side, there is clearly now much more information available than two years ago about the likely future of primary uranium supply. Although the number of so-called 'junior' uranium companies has reached over 400 and their plans are still very uncertain, many mines which were formerly at the 'prospective' stage have now much firmer planning and regulatory work behind them. The relative stability of primary production over the past few years, despite the significant uranium price increase, can be explained by a number of special factors but also the lengthy time lags in getting new production established, after so many years of depression and low investment in the sector.

The report evaluates all anticipated future mines and produces three scenarios. Even the lowest of these sees primary production rising from 40,000 tU today to 54,000 tU in 2010 and 63,000 tU in 2015. The upper scenario shows 65,000 tU in 2010 and 82,000 tU in 2015, with the reference scenario somewhere in between. The most important country is clearly Kazakhstan, which is set to become the world's leading producer in around 2010, taking over from Canada. Over the longer term, beyond 2020, it becomes harder to identify which of the many prospective mines will actually operate but BHP Billion's Olympic Dam mine in South Australia is clearly a key feature, with production likely to be increased from no more than 4,000 tU today to 12,000 tU and upwards beyond 2015.

Secondary supplies of uranium must also not be forgotten. Some commentators have tried to write these off as a thing of the past, but it's clear that this is far from the general truth. There will likely still be considerable quantities of highly enriched uranium (HEU) for downblending to civil reactor fuel beyond 2013, when the current 500 tonne deal between the United States and Russia expires. Most of this will be in Russia but the US Department of Energy itself has considerable stocks of surplus fissile material, notably some depleted uranium with relatively high tails assays, suitable for so-called re-enrichment to reactor assays at enrichment plants. So secondary supplies will remain an important part of the nuclear fuels market up to 2030, albeit at less relatively significant levels to the period from 1985 to date.

Bringing the demand and supply sides of the market in Chapter 5 is accomplished firstly by examining the Western and Russian sides of the market in isolation (given that they still remain segmented to a great degree) and then combining them, to show the overall world picture. The conclusions are quite clear. The uranium market should be well-supplied in the period to 2015 and indeed, given the strong upward movement in primary production, there could be notable surpluses recorded in several years. In the period thereafter, when demand is beginning to rise more rapidly in the reference and upper demand scenarios, there is a need for more of the identified prospective mines to come into production in order to satisfy demand (although some of the surpluses built up prior to 2015 could then potentially be run down). There are clearly more than enough uranium reserves already identified to accomplish this, while the current enhanced level of exploration activity is likely to yield additional good deposits amenable to mining in this timeframe. In the uranium conversion, enrichment and fuel fabrication markets, provided that the anticipated level of investment in new facilities is forthcoming, supply and demand should also be much in balance. The overall conclusion, therefore, is that there should be, in reality, no likely constraint on the global nuclear growth shown in the upper demand scenario by any nuclear fuel supply shortages.

The analysis of the new WNA report should be sobering for those who have incorrectly analysed the market in recent years. Although there have been real issues concerned with the apparent shortages in the uranium market since 2003, they do not presage any longer-term difficulty. Uranium is geologically abundant and the only issue has been to expand production sufficiently after a long period of depressed prospects. The uranium market, as it stands, has failed abysmally to communicate the correct signals to market participants in a timely manner, so prices had to rise very sharply to encourage new production. Demand will likely never be as high as the price bulls have predicted, while production will be much more flexible upwards than they have envisaged. Some people never learn the rules of economics – if you put up the price of something by an order of magnitude, you must have some influence on both demand and supply, even for a commodity where both are somewhat 'sticky'. If the WNA report achieves anything, it may be to encourage people to look at reality rather than their own particular dreamland.

Uranium: is there sufficient to fuel nuclear growth?

As nuclear power is discussed more intensively in the mass media, we have begun to hear once again an old argument that anti-nuclear people used to employ. This is that uranium is a very scarce commodity and its lack of availability will seriously constrain any significant expansion of nuclear power in the future.

If we go back to the period when nuclear power was expanding rapidly in the 1970s, it was indeed believed that uranium availability would be a real constraint. Reactor plans then envisaged saw world nuclear generating capacity rapidly rising to over 1000 GWe by the year 2000, at a time when proven Western uranium reserves had been heavily depleted by the uranium boom of the 1950s to fuel the nuclear weapons programmes. Back in the 1970s, fast breeder reactors were seen as the ultimate answer, but things never turned out as envisaged. Reactor programmes were cancelled to the extent that nuclear capacity by 2000 was only one third of what had been foreseen, while a renewed uranium exploration boom and subsequent mine development greatly improved the prospective uranium supply picture. Then secondary supplies of uranium, many originating from the earlier weapons programmes, became available on the market and mixed oxide (MOX) fuel, rather than fast breeder reactors, became the main user of separated plutonium.

This, in itself, should be a warning about the perils of making dire predictions of scarcity. There is a lesson in there somewhere about markets generally sorting things out somehow or other – necessity is also the mother of invention. Yet the nuclear industry must now take seriously the claims that uranium availability will constrain its future, particularly as we're now talking about ten-fold and more expansions of nuclear capacity being needed to fuel a hydrogen economy and other applications such as seawater desalination.

It is useful to split the scarcity argument into three time periods. The first covers the next ten years or so, when all industry analysts are agreed that primary uranium production now needs to expand rapidly to satisfy demand levels for nuclear fuel, which are predictable with some certainty, at a time when secondary supplies are dwindling. The second period goes well beyond this, for another 20-30 years or so, when we can foresee a serious expansion of nuclear power leading to a doubling – and maybe more – of demand for nuclear fuel. The third period covers the very long term, the remainder of this century and beyond, when technologies will likely change appreciably and the question becomes one of the extent to which uranium fits in with sound sustainable development principles.

For the first period, everyone accepts that there is more than enough uranium in proven reserves to satisfy demand. The question is one of the supply industry's ability to get the uranium out of the ground in a timely manner to satisfy steadily-rising market demand. World uranium production was just over 40,000 tonnes in 2007 and most analysts believe that this must rise by at least 50% over the next five to ten years as secondary supplies begin to dwindle and demand grows. Anti-nuclear voices, unfortunately assisted by careless scare-mongering from some industry analysts, assert that the uranium supply industry has been decimated by its past problems and now faces regulatory, financial and other challenges which will make it difficult to expand production to the extent required. Possible fuel shortages have even been mooted with reactors going short, with reloads and possible startups getting delayed.

This is all a mistaken fantasy. The uranium price has already reacted sharply upwards which is the signal to incentivise producers to expand production from existing mines as far as they can and also accelerate development of long-mooted mines from well-known deposits. It can certainly be argued that the uranium market has worked very imperfectly and that this price signal was delayed by several years from when it should have been sent, but it has come all the same. Producers certainly face challenges in obtaining regulatory approvals and mine developments are frequently delayed for a whole host of reasons, but it is reasonable to be optimistic that they will rise to the challenge over the next five to ten years. This is a lesson from the past – production is closely correlated with the uranium price. Why would one expect otherwise?

Some new producers are likely to appear in the next few years, stimulated by the market price increase, but most of the necessary expansion is likely to be led by today's leading producers, such as Cameco, Areva and BHP Billiton (the owner of the Olympic Dam mine in South Australia, which has huge potential for expansion). The rising national star in world uranium production is Kazakhstan, where joint ventures with Western, Russian and Asian companies are likely to lead to a significant increase in output over the next ten years, perhaps even putting the country ahead of Canada and Australia in the 'top three club' of big producers.

Turning to the period beyond this, stretching out into the 2020s and 2030s, there may well be a significant expansion of nuclear power, which will result in nuclear fuel requirements doubling or even trebling. This will be a dynamic period with lots of unpredictables, but critics take a very static view of uranium resource development. They take the proven uranium reserve level of today, at about 3.5 million tonnes, and claim that this will soon run out as gross annual reactor requirements are around 65,000 tonnes per annum, so there is only 50 years of supply at the current rate of use.

Yet uranium is not scarce in any geological sense, as its abundance in the Earth's crust is similar to that of tin, tungsten and molybdenum. Predictions of scarcity fail to take into account the likelihood of many new discoveries of uranium deposits. The exploration industry has already been greatly stimulated by the market price rise (up to 100 new junior companies have suddenly appeared) while exploration techniques are improving all the time. Mining technology is also improving, allowing access to deposits formerly judged too difficult to exploit and therefore uneconomic. For example, the development of in situ leaching (ISL) technology has allowed economic exploitation of some low-grade deposits. Finally, the doomsters have a bad understanding of mineral economics as a whole, where what is economic over time changes rapidly in line with market prices and technical advances.

The requirement for more uranium to fuel a nuclear 'boom' after 2015 is not going to come overnight. Indeed, the time horizons of the nuclear reactor constructors and uranium mine developers are identical. Both are long-term businesses and it should be possible to match up supply and demand very well, given appropriate price signals. Interest in uranium as a commodity was almost non-existent up to the start of the price rises four years ago, but now is intense and there is a reasonable expectation that the next period will see some valuable new discoveries. What is really needed, however, is a large number of new reactor orders in leading nuclear countries – *i.e.* confirmation that talk of a 'nuclear renaissance' is not just hype – to encourage potential producers to expand their interests further.

The nuclear industry can also point out that there are other possibilities in its favour. Secondary supplies are not yet 'dead' and will remain an important part of the nuclear fuel market for many years to come. A higher enrichment element can save a certain amount of uranium. Reprocessing of spent fuel and subsequent recycling of reprocessed uranium and plutonium in MOX fuel now looks increasingly economic as uranium prices go higher. This is clearly subject to a lot of political and other issues, such as priorities in spent fuel management, but is attracting increased interest (even in the United States), as expectations of nuclear fuel requirements increase.

It is therefore highly unlikely that lack of uranium will be a constraint on nuclear expansion in the period up to 2050. This is outlined in a World Nuclear Association Position Statement entitled *Can Uranium Supplies Sustain the Global Nuclear Renaissance?*

In the very long term, a whole range of new possibilities open up. In the earlier periods, the working assumption is that the reactor types in operation will be little different from those in operation today – essentially evolutionary light water reactors in the main. We can therefore be relatively certain on their fuelling requirements. Yet reactor technology is likely to change dramatically with the Generation IV reactors and beyond. These are all set to be much more efficient in their use of nuclear fuel – not just to save on cost but also to make used fuel management so much easier. Breeder reactors may well then come in, much later than was originally envisaged – as it was found that they were not needed before. Fuel types may also change appreciably – thorium could conceivably be developed as an alternative nuclear fuel.

When discussing nuclear in the context of sustainable development, it is important to emphasise these likely technological shifts. Uranium (and even thorium) resources in the world are clearly finite and in the very long term it is inevitable that a major nuclear expansion on today's technology would eventually begin to put them under pressure. Indeed rather similar to what is happening with oil today. So the industry must show that it can adapt in a way that shows it can, indeed, be sustainable and offer solutions for future generations. This is, of course, politically a very strongly charged area and nuclear faces particular challenges in being accepted. Demonstrating that it is still a relatively youthful industry with many possibilities of exciting technological developments would be a good start.

In conclusion, we may say that fears of potential uranium shortages are almost certainly misplaced, whatever time period is considered. Examining the short history of nuclear, combined with careful thought and analysis, suggests that this is one argument used by nuclear critics that just doesn't add up. Future uranium discoveries and mine development of course have lots of uncertainties surrounding them, but to over-emphasise these is foolish.

More uranium: when and from where?

The world uranium market continues to fascinate. Prices on the spot market are now at about $70 per pound (April 2008) after spending many years rooted at around the $10 level. Although many other metals and minerals are also experiencing rapid price escalation, the uranium situation has attracted an enormous amount of attention from people who previously had no interest whatsoever in either uranium or nuclear. This is very welcome for the industry, but may be relatively short term. Speculators, who will likely disappear as quickly as they came, have undoubtedly led some of the price escalation. The ranks of junior uranium companies (now numbering over 400 on some estimates) will thin through consolidation over time, particularly if prices fall back. Yet it is commonly agreed that the world now needs a significantly higher level of primary uranium production to make up for the diminution in secondary supplies (which have kept the market in balance for 20 years now) and rising demand for uranium from the nuclear power programmes of China, India, Russia and (hopefully too) the United States and United Kingdom.

There are, however, few obvious signs that more uranium is on its way, in the short term at least. World production of uranium has been around 40,000 tonnes annually in the entire period 2004-2007. One would surely expect, *a priori*, that higher prices should induce much more production. Supply curves in our economics text books normally slope upwards – at higher prices, producers should be willing to offer more goods, on the basis that this should be highly profitable for them. Mines previously 'out of the money', *i.e.* with production costs above the price level, should suddenly become active and their owners make good profits, assuming they can obtain contracts at the new, higher spot price level. Existing producers have every incentive to stretch production to the maximum, as each pound of uranium must be earning a much improved profit. So what is wrong? The answer is important as anti-nuclear people have begun to pick up on escalating uranium prices and some predictions of shortages to claim that there isn't enough uranium to sustain an upsurge in nuclear power and/or that it will be necessary to exploit increasingly poor grades in future, implying higher costs and carbon emissions from the fuel cycle. To them, uranium is really no different to oil, where the 'peak oil' proponents are now rubbing their hands that it is, at last, apparently beginning to run out.

The first point to make is that there is no shortage of uranium resources in the world. Uranium is abundant geologically and there is now a significant upturn in exploration underway, following the price hike. In any case, proven reserves are more than sufficient to fuel a significant expansion in nuclear power – beyond that, further economic resources will undoubtedly be discovered, while new reactor types are almost certain to economise significantly on the quantity of uranium required.

What is at issue, therefore, is turning these known resources of uranium into production. For those associated with existing operating mines, it is difficult to develop them more quickly. Operating mines have been running at high capacity utilisation factors for many years now, in an attempt to minimise production costs, so it is unrealistic to expect major increments to production from here. The limitation, as often as not, is milling capacity. Where this can be expanded, such as at the McArthur River mine in Saskatchewan, Canada, this is happening, but must await both full regulatory approvals and new investment. The message is that uranium production from these mines cannot expand very much, in the short term at least. One must also consider the position of the mine owners and operators. Battered by years of low prices, they would not be human if they didn't take the opportunity to bask in the current higher prices and simply take more profit from the existing quantity of production. Profits are now indeed growing significantly, as lower-priced long-term contracts gradually unwind, but these companies are unlikely to rest on their laurels. They will try to expand production and will undoubtedly succeed in doing so wherever they can. Having already put huge investment into such mines, incremental production may be achievable at modest capital cost.

One exceptional example is provided by the Olympic Dam mine in South Australia. This produces large quantities of both copper and uranium, with less significant quantities of gold and silver too. Current production capacity is around 4,000 tonnes of uranium per annum, but its owner (BHP Billiton) is looking at a major mine expansion, possibly increasing annual uranium production to 12,000 tonnes per annum and beyond. This is much more than the incremental expansions being considered elsewhere, but can only happen in the period beyond 2010, such is the scale of investment required.

It is, however, new mines that should bring the majority of the new production required to meet market demand. Yet these take time both to win the necessary approvals and also to be developed. Some constraints have appeared, such as the shortage of qualified mining engineers with uranium experience and the pressure from other mining ventures on key capital equipment such as large trucks and drilling equipment. These will be overcome in time, but it is clear that new uranium mines take a significant time to develop. The current generation of mines have reserves discovered 20 years and longer ago and new discoveries made today are likely to take similar timescales. So new mines coming into operation in the next few years are still likely to be based on well-known deposits, lying unexploited for many years.

Among the major national producers, Canada's output is likely to rise only slowly over the next five years. The Cigar Lake mine should eventually come into operation sometime after 2011, but much of its production will not be incremental as it will make use of existing milling facilities, already taking material from other orebodies which are becoming exhausted. Current operating mines can only expand slowly as milling capacities and regulatory limits are reached. Further new mines are someway in the distance, possibly resulting from the current surge in exploration activity.

In Australia. the most important development is likely to be the expansion at Olympic Dam already mentioned, but this will not happen for several years. In the shorter term, the Honeymoon in situ leaching (ISL) operation should start up, but there is nothing else on the immediate horizon. The easing of the anti-uranium mining sentiment which has occurred in official circles and the opening up of the possibility of uranium exports to China will take several years to have an impact on production, but are very positive features for the period beyond 2010. A possible cloud on the horizon is the difficulty of developing the Jabiluka deposit as a replacement for the Ranger mine as it runs out of reserves during this period. High prices are certainly an incentive, but approval of the local aboriginal landowners is required.

It is in Kazakhstan that the best prospect of substantial near-term production expansion exists. The Kazakhs have plans to expand uranium production from around 5,000 tonnes per annum at present to 15,000 tonnes by 2010-2012. This will be achieved by a mixture of joint venture operations with overseas partners (including Cameco, Areva, the Russians, Koreans, Japanese and Chinese) plus expanded production in mines wholly owned by the state producer Kazatomprom. All will be ISL mines and it is clear that the scale of this expansion will be very challenging but not impossible. As the new mines are likely to be low-cost (estimates of full costs – including capital – are around the $15 per pound level) the profit potential is substantial and uranium will become a major revenue earner and employment generator in the country.

The potential for near-term production expansion in the United States from ISL and other operations is also present, but will be of much lesser scale. Annual US production is likely to rise steadily over the next few years from its current level of only 1,500 tonnes, but it is unlikely to exceed 4,000 to 5,000 tonnes by 2010. Longer term, the potential is more substantial, assuming prices remain at something like the current level.

Elsewhere, the best prospects for higher production by 2010 are in Africa, particularly in Namibia and South Africa. Here the Langer Heinrich and Dominion mines have come into operation and should be followed by further new mines by 2010. Elsewhere, the future of the Rössing mine in Namibia has been secured for some years to come, while the operations in Niger could potentially be expanded if required.

Uranium production will therefore expand substantially by 2010. This may not, in any case, be sustained and new operations will be planned on the basis of survival at substantially lower prices. The upward trend in production should go some way towards satisfying those critics who predict continuous uranium shortages but the really big increases are likely to come only after 2010. The essential message is that it simply takes time to develop new mines. The uranium market has not worked well in the recent past as the necessary price incentive for increased production was continuously delayed by the appearance of more and more secondary supplies. Now there is the potential for it to function much better in the future.

Uranium: more production needed post 2013?

There is a common misconception that uranium is rather like diamonds, gold and other precious metals, which are only found in a few locations around the world. These are often in remote areas and the mineral deposits difficult to exploit – hence, with the scarcity, prices are high. Connecting uranium with these commodities is perhaps understandable, given the attention it receives in the media, emphasising its special properties. But it's fundamentally different, as in a geological sense it's quite abundant, with low concentrations observable throughout the Earth's land mass and in seawater too.

Students are surprised to be told that if they dig down underneath their lecture hall, they would likely find uranium in very low concentrations. The other important point is, of course, that, despite its abundance, it's only commercially viable to mine uranium in a small number of locations, where grades are particularly high or there are other advantages. Hence uranium production today is now concentrated in a small number of large operations in a limited number of countries. Talk of uranium scarcity by those opposed to nuclear power is really nonsensical – it's all around us but just needs a commercial incentive to exploit it. This has been lacking in the period when secondary supplies, from ex-military materials and inventories, have been an important part of the market but times have now changed, with prices up by a factor of seven over the past five years.

This is highlighted by the 2006 edition of the 'Red Book', the biennial publication on uranium resources, production and demand, produced jointly by the International Atomic Energy Agency (IAEA) and the Nuclear Energy Agency (NEA) of the OECD. This reviews the overall world picture in each of these areas but then contains extensive country reports on all the countries that have investigated their own uranium resources. It is certainly the 'bible' for the industry on uranium resources and highlights that exploration has already reacted sharply to the increased prices. It is too early to expect major new discoveries but re-evaluations of deposits already well-known from past exploration efforts have pushed up the low cost reserves total, (exploitable at less than $80 per kilogram). This now stands at 2.64 million tonnes, which at the current rate of use (about 65,000 tonnes of uranium per annum to fuel the 439 reactors around the world) would last for 40 years. This assumes all demand is met by primary production, but secondary supplies will remain an important part of the market for many years to come. Indeed, with the anticipated move back towards reprocessing used fuel and the prospect of Generation IV reactors, maybe in the late 2020s, secondary supplies may once again increase in importance. But, in any case, the low cost reserves are only the tip of a huge iceberg – the full resources which could conceivably become economic in the next 20 years are nearer 20 million tonnes, or ten times higher.

Any sensible concern about uranium's availability to fuel current and future reactors should therefore only surround the issue of actually getting the stuff out of the ground and to the market. The rapid price escalation has so far led to little in the way of production increases. Existing mines have been working flat out in recent years, while it simply takes time to develop the new facilities. New production may now be highly economic, but it can't be turned on and off like a tap.

In fact, it's important to recognise that uranium production facilities and their customers, the power reactors around the world, operate on very similar timeframes. It takes a long time to get new reactors up and running, but once they're in operation, they should run for 40-60 years. The licence extensions received for many existing reactors illustrate this point – nuclear economics supports keeping low-cost facilities open rather than decommissioning them. Uranium production is a similarly long-term business – mine lives may often be less than 40 years, but it takes many years to get new facilities in operation. Hence there should be no mismatch between the buyers and sellers of uranium – the reactor operators should be happy to grant the mine owners the long-term contracts they require to borrow money and develop their facilities. Prices in recent years were too low to allow this to happen, but now there seems to be a return to more sensible long-term contracting. It's possible that prices will eventually fall back, but hopefully not to levels which prevent the required new production from coming on stream.

Most projections show the world uranium market remaining tight for the next few years, until various production increases, from both existing and new mines, can reach the market. Reactors will certainly get their fuel, but the operators may have to pay much more for it than they expected a few years ago. This will have a very limited impact on the economics of nuclear reactors, but has been a blow to the professional pride of the fuel buyers. They would now no doubt concede that they were somewhat complacent and short-sighted about the market a few years ago, but the years of cheap and available supplies were so prolonged that it was perhaps understandable to believe they would last forever.

Beyond the shorter term (at least in nuclear terms), the period beyond 2013 is interesting for two reasons. Firstly, the Russians have announced that there will be no 'HEU-II' deal. In other words, the existing arrangement whereby downblended highly enriched uranium (HEU) from Russia reaches the world nuclear fuel market, will not be extended beyond 2013. Secondly, the upsurge in reactor orders, which is now expected in China, India and in some Western countries such as the United States and United Kingdom, should begin to have a substantial impact on the nuclear fuel market by then.

Although there still should be substantial quantities of HEU surplus to weapons programmes after 2013, it now seems that this will reach the commercial market more slowly than previously expected. The Russians feel that the HEU deal was seriously disadvantageous to them in commercial terms, effectively releasing a strategic asset on the world market far too quickly and cheaply. At the time of signing, they needed hard currency, but with strong oil and gas export revenues, this is no longer the case. The impact of the current HEU deal on the uranium market is up to 9,000 tonnes per annum, so equivalent to one very large mine. There will therefore be a potential gap created in the market when it ends, but given that the Russians will probably carry on downblending some HEU for their own purposes after 2013, it may be somewhat less than the full annual 9,000 tonnes.

Although some of the 'uranium bulls' trying to hype the share prices of junior uranium companies have talked up the demand side, it will take time for any rebirth of nuclear power to be reflected in new uranium demand. The earliest that new reactors could commence operation in the United States and United Kingdom is likely to be around 2015-2018. Given the lags in the fuel cycle, uranium for first cores of new reactors is needed about two years before commencing operation, so the impact may occur at just the same time as the existing HEU deal ends. Although it is possible that fuel buyers will sign contracts as soon as the reactors are ordered, especially if they feel that supplies may be tighter if they wait, the underlying market position would seem to suggest that a further sharp increase in primary uranium production will be required by about 2013.

Perhaps coincidentally, this is just when the major expansion planned for the Olympic Dam mine in South Australia may come on stream. This could represent a tripling of the output from the current 4,000 tonnes per annum, so would in itself replace the supply lost by the ending of the HEU deal. Yet the picture is obviously much more complex than this. In Australia itself, the 5,000 tonnes per annum Ranger mine will likely be out of reserves by then and it may not be possible to exploit the superb nearby Jabiluka deposit. Assuming that prices remain at current levels for the whole of the period from now until then, the uranium supply picture by 2013 will no doubt look very different from today. Many new producers could then be in operation while Kazakhstan has firm plans to become the world's leading producer, overtaking Canada and Australia, as early as 2010.

We can therefore see the likely expansion of uranium production over the next 15 years as a two-stage process. The first increase, which should occur up to 2010, will largely be a correction from under-production in the late 1990s and early years of this century. Underlying demand has been rising steadily but only slowly, by about 2% per year, so it can be seen as largely a change in the composition of supply. The signal that increased production would soon be necessary was continuously delayed, as the market didn't work as well as it really should have done. Now the incentive to produce more is clearly there, so the response will surely come. In the period after 2013, however, the production upsurge will also have both a supply cause (the ending of the HEU deal) but also occur because demand is beginning to rise more quickly. Existing reactors are set to run on and on, while there should also be an increasing number of new ones. But this is all dependent on getting those new reactor orders, and preferably soon too.

Junior uranium companies: at last something new?

Things usually move rather slowly in the nuclear world. New reactors take a long time to gain approvals and then build, while the technology has also moved along relatively slowly since the 1940s and 1950s, when change was dramatic. Similarly on the fuel supply side, new uranium mines and enrichment facilities usually take many years to get through the planning and approval stage and the corporate structures today bear at least some resemblance to the distant past.

In one area, however, there are signs of more dynamic change taking place. At the turn of the new century, the uranium supply industry had been battered and bruised by years of low prices, largely caused by abundant secondary supplies. Only a few companies survived this era, so production had become increasingly concentrated in a small number of low-cost mines run by big companies, in only a few countries around the world such as Canada and Australia. There were only a few companies, such as URI and Strathmore, which could be identified as 'junior uranium companies' – in other words, smaller companies involved in exploration, mine development and small-scale production.

The position today has changed dramatically. In just a few years, upwards of 400 companies have emerged which claim to have uranium amongst their interests. The perception that uranium production needs to rise sharply to cover both higher anticipated demand and falling secondary supplies, spread very quickly. This has been reinforced by the rapid upward movement of uranium prices, which ended 2007 at around $90 per pound, after being stuck at around the $10 level for the 20 years prior to 2003. This has encouraged financial backers to make money available and the companies, particularly in Toronto, Vancouver and Melbourne, have raised large sums. In fact, the increased interest in nuclear currently being experienced in financial circles started with uranium. It has only more recently spread to the possibility of new reactors, as major political leaders and environmentalists have come out in favour of nuclear as a contributor to preventing global warming and ensuring energy security of supply.

There are, however, plenty of cynics surrounding the junior uranium company phenomenon. These would claim that the companies are merely mining the financial markets in yet another speculative frenzy, taking advantage of the wall of money always actively seeking the next big story. Uranium has suddenly become big news and they view the speculative money feeding at least part of the uranium price rise as another element in the process.

There has to be at least some truth in this – there has been at least an element of hype. For 'old hands' in the industry, some of the stories put out by many of the companies produce wry smiles of amusement. Reading some of the descriptions of new nuclear power programmes would lead one to believe that 100 new reactors will be urgently requiring massive uranium supplies within a few years. The names of hundreds of old known deposits have resurfaced, as have many of the long-retired people, now unexpectedly attracted out of retirement, who discovered and started to develop them. Credibility with the financiers is achieved by a good story involving previous exploration records combined with some people with previous 'form' in uranium.

It is notable, however, that the majority of the companies were previously involved in exploring for other metals and minerals, particularly gold, and have now added uranium to their portfolio. It's hard not to be rather cynical about this – once interest wanes and something else becomes 'flavour of the month', they are more than likely to head off in that direction. The Vancouver market, in particular, is populated by large numbers of such venture mining companies. Investors, however, tend to be aware that such companies are not for 'widows and orphans', but they provide a good gamble for those seeking higher than average returns.

In practice, many investors have done very well out of the junior companies, particularly those who got in early. With the uranium price rising steadily since 2003, share prices also tended to move upwards, although some have been very volatile and there have been 'market corrections' when general commodity prices have been on the slide. Prices have fallen back sharply, however, after the uranium price peaked in mid 2007.

What more do we know about the junior uranium companies? Firstly, it is clear that nearly all of them are involved at the early stage of exploration. It is believed that world expenditure on uranium exploration quadrupled between 2001 and 2006 and now stands at over $200 million per year. The juniors are responsible for over half of this, the remainder being carried out by the current major producing companies. Altogether, this can be called the beginning of a second major exploration cycle for uranium, following the first in the 1970s/1980s, when many of the current operating mines were discovered. Maybe a third of the companies are also buying up previously known deposits – indeed there has been a rush to acquire the data from previous exploration activities, while share swaps and other similar corporate activity is very much part of life for such companies. Very few of the companies, no more than 20, have yet reached the next stage, which can be termed active mine development, in other words going through the regulatory process, preparing environmental impact assessments and bankable feasibility studies and beginning to invest in the mine infrastructure. What is clear is that it will still take many years for new discoveries to result in active mine production – this is true for all metals and minerals, but in uranium it tends to be prolonged. In some cases, this may be 20 years or so – it has taken that long to get current mines into operation. Of the junior companies, only a small number are now producing, such as URI and Mestena in the United States, Paladin in Namibia and Uranium One in South Africa and Kazakhstan.

As time goes on, however, more and more junior companies will reach the production stage. It's interesting to consider the countries where this is likely to take place. About half the junior companies have their headquarters and stock exchange listing in Canada, with about one third in Australia. Yet the geographical spread of their activities is much greater. Although Canada and Australia are both prominent at the exploration and known deposit acquisition stages, many of the companies are active in both the United States and Africa, with smaller numbers working in Asia, Latin America and Europe. So new production could come from any of these locations – indeed there are good grounds to believe that Africa and United States may well outpace Canada and Australia, at least over the next 5-10 years.

With the exception of the Honeymoon ISL mine in South Australia (which Uranium One is expected to commission in 2008), there is nothing immediately coming up from the juniors in the two big producing countries, which currently account for almost half of world output. One reason for this may be that the regulatory process seems to be rather more lengthy in these countries (and Australia is only now escaping from anti-uranium public policy and sentiment). It is notable that Paladin have got the Langer Heinrich mine in Namibia up and running in a relatively short period of time, encouraged by supportive public authorities. Other African mines, in South Africa and other countries, are likely to get into production well before all the projects in Canada and Australia that are currently only being talked about. The position in the United States is less clear, as the regulatory process there can also be lengthy, but there is every prospect of some of the juniors getting production moving strongly upwards in the 5-10 year timeframe. American utilities, who have been most exposed to the uranium price spike, will no doubt support this. Yet the biggest increase in world production before 2010 is likely to come from Kazakhstan, where operations are largely controlled by Kazatomprom, clearly a major established player in the market.

Another feature to watch with the junior uranium companies is that they are often very dependent on the uranium price staying high. The average grades of many of their deposits are quite low, suggesting they may be quite high up the cost curve. While buyers are keen to diversify their sources of supply, particularly at present when they feel weak against the established producers, price is vitally important and they won't generally support those with high cost profiles.

It is clear that there will inevitably be a huge amount of consolidation in the junior uranium sector. The better companies will inevitably become acquisition targets for the major producers, particularly if these find themselves short of material to satisfy contracts. Indeed, cynics about the sector would claim that this is all most of the juniors really want to do – they have no intention of ever producing a pound of uranium themselves and are just seeking to sell out at a good profit to whoever comes along. This may be true of many, but there are clearly those who are dedicated to getting mines up and running. But of course, everything has its price. Over the longer term, there will likely be only 10-20 survivors out of the current pool of 400 who are still involved in uranium, independently and at the mine production stage. Possibly even fewer. So a wave of takeovers, mergers and even corporate failures is now likely to ensue.

Finally, it must be said that the junior companies have brought a fresh air to a sector that, with a few exceptions, appeared on its last legs only a few years ago. Some would say that a lot of the fresh air is also hot air, but this is inevitable in a business such as mining, with its sharp up and down swings. Industry meetings have certainly been enlivened by the appearance of many 'characters' from the past and by the arrival of financial types, only really out to make money (and not apologetic about this). And as said earlier, the rising interest in uranium has presaged the increased focus on nuclear power as a whole, which must be a good thing.

Enriching: is it more than a proliferation risk?

Much of the discussion of uranium enrichment over the past few years has been focused on the technology. The old, huge gas diffusion plants are clearly coming to the end of their lives and the question has been: what next? This discussion has now been resolved, for the time being at least, in favour of gas centrifuge technology, as already used on a commercial scale by Urenco and the Russians. Laser enrichment is retained as a possibility for the future but achieving substantial capacity at low operating costs has so far proved difficult.

More recently, however, some new issues have come up which are relevant to the enrichment sector. Firstly, the recent rise in the uranium price means that there should be an increased demand for enrichment to produce nuclear fuel. This is because utilities have flexibility in choosing the relative amounts of uranium and enrichment to commit to produce enriched uranium – if one component becomes more expensive, it is logical to use more of the other. Secondly, much recent attention has focused on the quality of nuclear industry supply infrastructure, with the sudden realisation that it is perhaps rather more fragile than we thought. In the enrichment sector, however, there are huge sums being committed to new facilities in the United States and France, in addition to the continuing programme of investment at the existing gas centrifuge plants. Finally there are all the nuclear technology proliferation concerns, which have cropped up in recent years. Many of these are focused on uranium enrichment and have led to popular media interest in what was formerly regarded as a rather obscure subject, for the eyes of specialists only. The proposals to deal with these concerns are highly controversial and will undoubtedly have some effect on the commercial nuclear sector.

It's clear that the increased uranium price, assuming it is sustained, will push up world enrichment demand. This is because in seeking a quantity of enriched uranium to fabricate as fuel, utilities can choose an optimum contractual tails assay, a commercial decision based on relative uranium and enrichment prices. The U-235 assay of the tails (depleted uranium) depends on the amount of enrichment services (SWUs) applied to the natural uranium, which was supplied as the feedstock for the process of producing enriched uranium. During the period of low uranium prices, there was a significant difference between the tails assays typical for Western and Russian enrichment plants. In the West, low uranium prices pushed the optimum tails assay up to around 0.35%, as it became more economical for utilities to supply a greater quantity of uranium and use fewer SWUs. In Russia, however, substantial spare enrichment capacity and an increasing shortage of natural uranium pushed tails assays down to as low as 0.10%, as stockpiles of depleted uranium (including material from the West) have been sent through the enrichment plants once again.

With the substantial increase in world uranium prices, the optimum tails assay for Western utilities has now moved down from 0.35% U-235 to below 0.25%. This has had a major influence on enrichment demand – in the United States alone, a movement from 0.35% to 0.25% would increase enrichment requirements by more than 2 million SWUs per annum, but reduce uranium demand by around 4000 tU per annum.

This has clear implications for commercial enrichment facilities. It is clear that a major investment programme in these is long overdue. The large gas diffusion plants in the United States and France are now very old and are very inefficient in their use of electricity, while it is clear that gas centrifuges are the preferred replacement technology.

Indeed, it has only been the arrival on the world market of a substantial number of SWUs from downblended Russian HEU which allowed the heavy investment programme to be delayed. It can be argued that only this, plus surplus enrichment capacity in Russia, prevented the supply problems apparent in other areas of the fuel cycle. For example, disruptions at large world uranium mines, such as McArthur River, Olympic Dam and Ranger and the realisation that the primary uranium sector had been severely battered by financial pressures for many years, were major triggers for the sharp increase in uranium prices since 2003.

In the enrichment sector, nevertheless, over $10 billion is being invested in three new facilities, two in the United States and one in France. In the United States, with the American Centrifuge Plant, USEC has revitalised the old Department of Energy technology and aims to open a new 3.5 million SWUs per annum (initially) plant at Portsmouth, Ohio. Urenco and its partners in the National Enrichment Facility plan to open a centrifuge plant of 3 million SWUs per annum in New Mexico. Finally, Eurodif in France will replace its old Georges Besse gas diffusion plant with a new centrifuge plant of 7.5 million SWUs per annum capacity, based on technology-sharing with Urenco. In addition, Areva has also announced its intention to build a new plant in the United States.

These new facilities are a mixture of replacement and incremental capacity (USEC and Eurodif are essentially more efficient replacements but the New Mexico facility is incremental and gives US and other utilities another supply option). The new investment is very welcome but getting the new plants running well and on time will be very challenging. For the two US plants, in addition to the technical challenges, the Nuclear Regulatory Commission (NRC) has a very heavy workload for licensing, with Yucca Mountain, nuclear plant up-rates, relicensing and now new reactor build applications all coming along at the same time as new enrichment facilities.

Any shortfall in Western capacity can theoretically be made up by the surplus in Russia. It is believed that the Russian gas centrifuge capacity is already up to 25 million SWUs per annum but it is expected that this will rise slowly as old centrifuges are replaced by new improved models. It is clear that Russia now wishes to sell higher value nuclear services to the West, rather than being seen as a cheap fuel supplier; its enrichment capacity is a key to this. The strategy is sometimes referred to as 'SWU for U' in the sense that a large amount of the capacity is now being used to enrich tails material, both sent to Russia from Western enrichment plants but also domestically-sourced. This is providing an additional source of nuclear fuel. However, the Russians would eventually like to supply their enrichment services directly to all Western utilities, something currently not possible owing to trade restrictions.

Finally, turning to the renewed proliferation concerns on enrichment, these essentially started with the announcement from North Korea claiming it had an operating centrifuge enrichment programme. There were substantial doubts about the status of this but it was followed by further revelations from Iran and Libya, showing that they were developing similar programmes. Centrifuge enrichment technology is very difficult to master and needs high-quality plant components, but it appears that in each case, some progress had been made towards achieving facilities which could enrich uranium to weapons-level assays.

The common link in each of these countries has been technology transfer from the enrichment programme in Pakistan, which uses old Urenco-derived centrifuge technology. This has clearly worried those concerned with weapons proliferation, although the quantities of enriched material produced and its assays remain unknown. Then Brazil added to the worries by initially refusing to accept full International Atomic Energy Agency inspections at its enrichment facility at Resende. The reason given for this was advanced knowledge-protection (the Brazilian programme is essentially independent) but suspicions persist that there are hidden weapons proliferation intentions.

These revelations have led to proposals for strengthening the non-proliferation regime. In addition to the need to accelerate adherence to the IAEA Additional Protocol, which ensures a stricter inspection regime, the United States and others have proposed that enrichment facilities should be confined to the small number of countries already involved in the business. These will then offer full and fair trade to only those who accept full scope safeguards. A similar regime is proposed for spent fuel reprocessing, which also carries proliferation risks.

The idea of limiting enrichment and reprocessing facilities to a small number of countries has a good and bad side in a commercial sense. Such facilities need economies of scale and having them spread over many countries must imply higher costs. The downside is that such proposals could also potentially serve to limit competition and push up prices for buyers. Some commentators certainly see the US proposals as having a commercial slant – memories are long in the industry, going back to when the United States held a stranglehold over fuel supply through its monopoly supply of enrichment services.

The other important point is that the civil nuclear fuel cycle has never been a significant contributor to weapons proliferation, so constraining it in various ways will not get to the root of the problem. Nations seeking nuclear weapons have taken more direct routes, through research reactor fuel reprocessing or direct acquisition of enrichment technology, without any civil justification.

It must therefore be hoped that whatever arrangements are brought in to meet proliferation objectives do not diminish competition in this important sector of the fuel cycle and allow nuclear utilities around the world to secure regular and timely supplies of enrichment services at keen prices.

Additional enrichment: a substitute for expensive uranium?

With uranium prices in April 2008 at around $70 per pound (after a long period around the $10 level), there is clearly every incentive for reactor operators to economise on its use. However, the extent to which it is possible for buyers to save money by using more enrichment is worthy of close examination, as it is not as simple a matter as it may at first seem.

Reactor operators require enriched fuel to supply to their fuel fabricator and can choose a variable mixture of uranium and enrichment services (*i.e.* SWUs) to achieve this. This choice is made through deciding on the contractual tails assay to be used at the enrichment plant. Natural uranium has a U-235 assay of 0.71% and the amount of SWUs applied to the feed material determines the assay of the waste stream (*i.e.* the tails assay). If the buyer chooses a low tails assay (perhaps 0.25% in present conditions), he can supply a smaller amount of uranium as feed to the enrichment company, which will then apply more SWUs to obtain the required quantity of enriched uranium. Alternatively, in choosing a high tails assay (say 0.35%), the fuel buyer will supply a larger quantity of feed uranium, which will then receive fewer SWUs to produce an identical amount of enriched uranium.

The decision on the contractual tails assay is motivated by the relative prices of uranium and enrichment services and for each price relationship, there is an optimum rate. As the price of uranium has risen (with the price of SWUs remaining relatively stable), the optimum tails assay has fallen from around 0.35% towards 0.20%. In theory, a reduction of the tails assay of this magnitude could reduce uranium requirements by up to 25%, while increasing the demand for SWUs by a similar amount. There are, however, good reasons to expect the effect to be rather smaller than this, as the rather theoretical basic calculation contains a number of underlying assumptions.

The most obvious of these is that it assumes that there is sufficient spare enrichment capacity available to supply the additional SWUs that will be required. For the Western enrichment companies, this is certainly doubtful. The part of current capacity represented by the old gaseous diffusion plants is gradually being closed down and both USEC in the United States and Eurodif in France are moving towards establishing new centrifuge facilities. Urenco in Europe, which has been using centrifuges for many years, has been steadily increasing capacity to match the contracts it has received but does not maintain any spare capacity to cope with sudden fluctuations in demand. It will be some years before its incremental capacity in its new US plant becomes available.

On the other hand, it has always been assumed that the Russian enrichment plants, with rated capacity of 25 million SWUs per annum, have a lot of spare capacity. These plants are subject to trade restrictions from the West, limiting the extent to which they can contract with Western buyers, but, more importantly, the magnitude of their spare capacity is now also somewhat in doubt. Making realistic assumptions about the very low tails assays (down to 0.10%) that they use and their needs to fulfil the requirements of Russian-designed reactors worldwide, contracts with Western utilities, the needs for downblending Russian highly enriched uranium (HEU) for the contract with the United States and also for re-enriching old tails material in stockpiles, it is possible to account for all of the 25 million SWUs capacity each year.

If capacity is tight, the obvious impact of buyers requesting lower contractual tails assays will be for the price of SWUs to rise. Once this happens, the incentive for lower tails assays will diminish – indeed, if the enrichment price rises substantially, it may be completely eliminated. From the point of view of the enrichment companies, they are unlikely to go about installing new capacity solely to cope with a reduction in the optimum tails assay, which may well be only temporary. Or if they do, they will want to receive substantially higher prices for the additional SWUs, therefore diminishing the attraction of switching.

Given that there are already substantial new investments taking place in new centrifuge enrichment plants, the future enrichment price is also very hard to assess. Centrifuge enrichment uses much less energy than the old diffusion technology, but billions of dollars of new capital are required to build the plants. The new capacity (apart from that of Urenco in the United States) is replacing facilities where the investments were amortised years ago and so must earn a good return on the new capital employed. So enrichment prices may well have to rise anyway, in addition to any effect from the pressure on capacity caused by the move towards lower tails assays. Indeed, it is conceivable to pose a scenario for the future where uranium prices have fallen back substantially from current levels, as additional low-cost production facilities come online, but where enrichment prices are substantially higher than today owing to the need to finance the new facilities. The impact of this, of course, would be to push optimal tails assays back up again.

In addition to the question of adequate capacity, enrichment companies have also become accustomed, in the recent past, to using a lower operating tails assay than that implied in contracts, a process known as 'underfeeding'. This generates additional uranium for their use and works as follows. The customer supplies a quantity of uranium as feed to the enrichment company, based on the contractual tails assay (which in turn is optimised from the uranium-SWU price relationship faced by the buyer). The relevant optimal tails assay for the enrichment company may well be different, as the marginal cost of using some more SWUs is probably substantially lower than their market price. This implies that the enrichment company may have a lower optimum tails assay and should therefore use more SWUs (and less uranium) than is implied in the contract with the customer – *i.e.* the operating tails assay is lower than the contractual. This allows the enrichment company to hold back some of the uranium supplied as feed and then sell it on the market as natural uranium or enrich it to sell as enriched uranium product (EUP). Note that the enrichment company has in no way 'stolen' the material – its obligation was only to supply a quantity of enriched uranium to its customer - how it produces this is its own business.

The impact of past underfeeding has been to soak up much spare enrichment capacity available after contracts have all been signed with customers. This makes it unlikely that there will be much spare capacity available to satisfy increased SWU demand flowing from higher uranium prices, at least in the short term.

The process can work in reverse, of course (which is called 'overfeeding'), where the enrichment company faces high marginal SWU production costs for some reason (maybe needing to install additional capacity just to cover one contract) and chooses to use additional uranium to that supplied by customers in order to produce the enriched uranium. This is rather unlikely in current market conditions, but demonstrates the point that enrichment plants and their customers do not necessarily face the same optimal tails assay – indeed, it is most likely that they won't.

A final consideration concerns the recent practice of Western enrichment companies sending the tails material produced at their plants each year to Russia to be re-enriched there to create additional enriched uranium. The economics of this are crucially dependent on the assay of the tails. If this is substantially lower than the assumed 0.30%, the financial advantages will rapidly diminish, cutting off a potential source of revenue for enrichment companies. Again the impact is to provide less incentive for enrichment companies to employ lower tails assays.

The pure arithmetic would suggest that the recent upsurge in uranium prices may theoretically cause a 25% reduction in annual uranium requirements (therefore a reduction of about 15,000 tonnes per annum), with a similar percentage increase in SWU requirements (or about 10 million SWUs per annum). What, however, will be the likely impact in the real world? The factors already mentioned suggest that it will be substantially lower than this, perhaps of the order of a reduction of uranium demand of 6,000 tonnes per annum and an increase in SWU demand of 3 million SWUs per annum.

There are definitely indications from world utilities that tails assays are beginning to fall (towards 0.20% in some cases) with the sharply higher uranium price, but it will also take time for the higher prices to feed through into current contracts. So a notable lag will be experienced, reinforced by one final complicating factor. The quantity of secondary uranium and SWU supplies still available in the market, not so affected by current price relativities, is also likely to slow things up. In 2007, around 40% of world uranium requirements were still satisfied by secondary supplies of one kind or another. These are influenced by rather different economic factors to those of primary supply sources. Yet once primary uranium production begins again to take a higher share of reactor requirements, the impact of varying tails assays should once again become much more transparent.

Enrichment: as interesting as uranium?

The international uranium market has received a significant amount of renewed attention over the past few years. Yet there is also a lot now happening in the enrichment sector as it undergoes a substantial technology change with much new investment, while some of the historic trading relationships in the market may also be set for a change.

There is, of course a link between uranium and enrichment supply, to the extent that they are at least partial substitutes. In order to obtain supplies of enriched uranium, required for nearly all commercial nuclear reactors, fuel buyers can alter the quantities of uranium and enrichment services by varying the contractual tails assay at the enrichment plant. When uranium becomes relatively more expensive, there is an incentive to supply less of this and use more enrichment, thus 'extracting' more U-235 from each pound. When uranium prices were around $10 per pound, the optimum tails assay was about 0.35% but with the dramatic increase in prices and only a minor upward movement of uranium enrichment prices, the optimum is now below 0.25%. Assuming such price relativities are sustained into the long term (which is arguable), there could be a substantial (20% and above) increase in enrichment demand and a corresponding fall in the requirements for fresh uranium. The major limitation on this is the availability of surplus enrichment capacity – constraints on this have so far limited the possibility of buyers to take full advantage. Nevertheless, much higher uranium prices are undoubtedly a positive feature for future enrichment demand and may be taken into account in the coming major plant investment decisions.

On the enrichment supply side, the most obvious feature is the gradual replacement of the old gas diffusion facilities of USEC in the United States and Eurodif in France with more modern and economical centrifuge plants. Even with favourable supply contracts, the huge amount of power required by the diffusion process renders it uneconomic against the centrifuges, as currently used by Urenco in Europe and by the Russian plants. Eurodif will gradually replace its capacity with centrifuges derived from a technology-sharing agreement with Urenco, while USEC has decided to develop its American

Centrifuge technology, based on Department of Energy programmes in the 1970s and 1980s. Assuming USEC can overcome the financing and technical issues surrounding its plans, the last gas diffusion capacity should disappear around 2015 and the whole of the enrichment market should then be covered by centrifuges. The only likely alternative is the Australian SILEX laser enrichment technology, which has gained the support of GE for its possible commercial development. Lasers may yet turn out to be the technology of the future, as was thought ten years ago when USEC and others were investing significant amounts in laser technology, but its widespread commercialisation (if it turns out to be technically and economically viable) may have to await the next generation of heavy investment in capacity, in the period after 2015. For the near future at least, centrifuges will be the technology of choice.

Urenco has plans to continue increasing its capacity in Europe on a modular basis to 11 million SWUs per year and beyond. In addition, in mid 2006 it achieved licensing approval for its National Enrichment Facility in New Mexico, which will eventually reach 3 million SWUs per year – much of this capacity has already been tied up in long-term sales contracts with US utilities, indicating that they very much welcome the arrival of a new source of US-based supply on the market. The Russian centrifuge capacity is not known with any degree of accuracy, but is likely to be in the range of 25 million SWUs per year. This is believed to be continuously rising, as old centrifuges are replaced by new.

On the commercial side, the key anticipated developments mainly surround Russia. After a period of much speculation, the Russians have announced that there will be no 'HEU-II' deal with the West after the current one expires in 2013. This has been supplying roughly half of the US enrichment requirements since its commencement in the mid 1990s and has also substantially contributed to important non-proliferation goals. The commercial terms, however, are judged by the Russians to be non-favourable, as they are effectively supplying USEC at wholesale prices when they would much rather supply the US market on more normal commercial terms. Russia needed hard currency at the time the deal was signed but now has strong oil and gas export earnings – hence their new strategy of playing a longer-term game with their nuclear fuel assets.

Although there should still continue to be substantial quantities of surplus Russian highly enriched uranium (HEU) available for down-blending in the period beyond 2013, it is now reasonable to expect that it will be mostly consumed by internal needs, to fuel Russian-origin reactors both at home and in export markets such as China and India. Some Russian enrichment capacity will therefore continue to be required to provide the blendstock for this, from re-enriching tails material. Yet it is clear that the Russians would like to use much more of their enrichment capacity to supply enrichment services to Western buyers on a direct basis, rather than use it for re-enriching tails sourced either domestically or from Eurodif and Urenco. This desire is currently constrained by trade restrictions in both the United States and Europe, but these are gradually melting away following new agreements and may, eventually, completely disappear. In this they have the support of many Western buyers, who see their interests as best-served by more competition in the market. To some extent, they will get more choice from the substantial investments now being made in Western centrifuge facilities, but access to the extensive Russian facilities should promote even more price competition in the market, at a time when they are paying more and more for their uranium.

There is also a connection here with the Russian desire to be a strong regional fuel cycle specialist in key areas such as enrichment and used fuel management, along the lines also proposed by Mohamed ElBaradei, the Director General of the International Atomic Energy Agency (IAEA), and by the United States under the Global Nuclear Energy Partnership (GNEP) programme. Although these are mainly aimed at securing non-proliferation goals, it may also make sound economic sense to have enrichment concentrated at a limited number of facilities worldwide. Certainly it makes good sense to use the available Russian capacity as fully as possible, provided it fits in with national and commercial security of supply objectives. Using it for re-enriching depleted uranium is very much a second-best alternative and would have dubious economics in the Western world, as it eventually must encounter rapidly diminishing returns as the assays of the tails material gets lower and lower.

Another desire of the Russians is to increase the market share of their enrichment services in Asian markets. It is clear that they have been at least partially successful in this, as they have now secured contracts with several prominent Asian utilities for part of their business and also opened sales offices in this region. It is not clear how the increasing level of demand in China, from its notable new reactor-building programme, will be satisfied. The existing enrichment facility at Lanzhou, which has both Russian-origin gas diffusion and centrifuge capacity, could be increased in capacity, but rising requirements may alternatively be sourced via imports from any of the overseas vendors. There has been a significant amount of news and comment about China's rising uranium requirements, but little on how they will source their future enrichment needs.

Another commercial issue in the enrichment market, the USEC antidumping case filed against both Eurodif and Urenco, is now of much less significance. The ruling that enrichment is a service rather than a product, meaning it can't be subject to duties, remains contentious but the magnitude of the assessed duties has now anyway diminished to the point that they are unlikely to have a significant impact on the market. With the enrichment price now having risen by over 50% from the lows of $80 per SWU at the time the case was first filed, the action has arguably succeeded to some extent, although the market shares of both Eurodif and Urenco in the US market have continued to rise.

In conclusion, there is certainly a great deal happening in the enrichment sector, which should keep it very much in the news. Demand should increase steadily as new nuclear power plants come into operation worldwide while the flexibility of the tails assay with higher uranium prices provides some additional interest. The new plant investments to replace old and inefficient capacity are long-overdue and would probably have been made some years ago if the nuclear fuel market had been stronger. The commercial arrangements in the market are also currently very dynamic, with the prospect of more competition for available business, especially in the United States. Finally, as a key sensitive area for the non-proliferation of nuclear weapons, the enrichment sector is likely to be a central point in the new international arrangements which must be developed to support a buoyant nuclear sector throughout this century.

Fuel fabrication: a different sort of market?

One of the peculiarities of the nuclear fuel cycle is the way in which utilities with nuclear power plants buy their fuel. Instead of buying fuel bundles from the fabricator, the usual approach is to buy uranium from the mines and then sign contracts with conversion, enrichment and fuel fabrication companies to eventually obtain the fuel in a form that can be loaded into the reactor. During presentations on nuclear power, financial companies and journalists frequently ask: "Why don't they just buy the fuel direct – why complicate things with all those separate contracts?"

This is a good question and the answer is partly to do with the history of how the fuel market developed. It is, however, more closely related today to two factors. Firstly, the financial risks a fabrication company would take on by having to take positions on uranium, conversion and enrichment price levels (memories of the Westinghouse case from the late 1970s, involving a major default when uranium prices rose sharply, are still prevalent). The fuel managers today take on these risks. Yet they seem happy to do so. Their choices are the second reason for continuation of the traditional approach. They believe – rightly or wrongly – that this offers them the best price and service. They will typically retain 2-3 suppliers for each stage of the fuel cycle, who compete for their business by tender.

On the face of it, little similarity exists between the workings of the uranium, conversion and enrichment markets and that of fuel fabrication. The first three are fungible – *i.e.* bulk commodities or services, with no product differentiation save for preferences on origin that exist in some markets. A pound of uranium is a pound of uranium and a SWU is a SWU. Nuclear fuel assemblies, on the other hand, are highly engineered products, made specially to each customer's individual specifications. These are determined by the physical characteristics of the reactor, and by the fuel cycle management strategy of the utility concerned. Despite this, fabrication costs as a share of total fuel costs are not so significant – typically no more than 20% of the total, rather lower than the share of both uranium and enrichment services.

Fabrication requirements are affected by changes in utilities' reactor operating and fuel management strategies, which are partly driven by technical improvements in fuel fabrication itself. For example, light water reactor (LWR) discharge burn-ups have steadily increased as improvements in fuel design have made this possible – this has tended to reduce fabrication demand, with fuel often now remaining in the reactor for a longer period. Longer refuelling cycles can also reduce fabrication demand. Fuel fabricators have had to adapt to these changes, driven by the strong competitive pressures in the LWR fabrication market.

Annual world requirements for LWR fuel fabrication worldwide are about 7000 tonnes of enriched uranium per annum. Requirements for CANDUs and other reactor types account for an additional 2000-3000 tonnes per year. Where fuel contains enriched uranium, annual requirements are only a fraction of annual natural uranium requirements (because most of the mass of this remains in the enrichment tails). Where natural uranium is used directly, as in CANDU and Magnox reactors, annual fabrication requirements are identical to uranium requirements.

Current Western fuel fabrication capacity is about 15,000 tonnes per year – Russia's is over 3000 tonnes per year. It is therefore clear that there is substantial overcapacity in fuel fabrication. In the

Western world alone, fabrication capacity outweighs requirements by some 40%. This has been the position for many years. Despite much rationalisation around the world, national facilities have been maintained for non-economic reasons, and this has perpetuated this feature. Yet where competition is allowed to prevail in fuel fabrication, the market is very competitive.

Many fuel fabricators are also reactor vendors, and they usually supplied the initial cores and early reloads for reactors built to their own designs. As the market developed, however, each fabricator began to offer reloads for its competitors' reactor designs. This led to the market for LWR fuel (PWRs in particular) becoming increasingly competitive. With several suppliers competing to supply virtually every different fuel design, a trend of continuous fuel design improvements has emerged. Traditionally the market for BWR fuel has not been as segmented and competitive. This is changing, however, as fabricators begin to focus more attention on this relatively less competitive market as a means to acquire additional market share.

Price is clearly a vital competitive tool in the fabrication market. In particular, the weakness of the US dollar against the Euro and Yen has recently increased the competitiveness of US vendors like Westinghouse and GE. Detailed contract terms are, however, just as important, as is the technical support a fabricator can offer. This is increasingly important as users try to get more out of the fuel in terms of burn-up – fuel failures are very costly.

Given the very competitive nature of the LWR fabrication business and the clear overcapacity in supply, the industry has reorganised in recent years with substantial corporate consolidation. BNFL acquired the nuclear business of Westinghouse of the USA, added ABB's nuclear operations to it, before selling it to Toshiba. Framatome and Siemens also merged their nuclear operations to form Framatome ANP, part of the Areva Group. Finally, General Electric formed Global Nuclear Fuels with its Japanese partners Toshiba and Hitachi. These moves may eventually lead to capacity reductions when production is consolidated at a smaller number of locations.

This reorganisation and consolidation to gain market share appears set to continue for some time yet. As this proceeds and operations are made more efficient or perhaps, in some cases are closed, the supply-demand balance of the market may begin to move closer to equilibrium. The intense competition in the PWR market will likely continue and spread more generally to the BWR market.

Outside the LWR fuel market, fuel fabrication requirements tend to be filled by facilities dedicated to one specific fuel design, usually operated by a domestic supplier. For example, all fabrication requirements for UK Magnox and AGR reactors are supplied by dedicated domestic facilities. Similarly, all Soviet-designed RBMK and VVER reactors have historically been supplied by dedicated facilities within Russia (although other fabricators have begun to offer to produce fuel for VVERs, especially those in countries which have now joined the European Union).

CANDU fuel is also produced almost exclusively within the country where the reactor is located, by facilities dedicated to such supply. The market is somewhat more diversified than the other non-LWR fabrication markets, so as a result, the requirements for some utilities with heavy water reactors may, on occasion, be met from a non-domestic supply source.

One area of the fabrication market where growth remains a possibility is in mixed oxide (MOX) fuel fabrication. Existing plans by the limited number of countries that have to date committed to using MOX fuel will require the expansion of capacity at existing MOX fuel fabrication facilities, along with the construction of some new facilities. Yet decisions made on MOX facilities are heavily political, in contrast to the more commercial decisions made in the LWR market.

What would really stimulate the fabrication market is, of course, some new reactor orders. The two EPRs in Europe, in Finland and France, will require first cores, which will be good incremental business. More new plant orders would be very welcome and help to remove the surplus capacity apparent in the market. The boom in Chinese reactor construction is not directly helpful to the Western vendors as the Chinese seek to maximise local content of fuel supply. With poor uranium resources, they are intent on expanding their domestic fuel fabrication facilities. The Western vendors will help set these up, but are unlikely to get much further business.

Fuel fabrication is clearly a crucial element in the fuel cycle and any disruption will impact directly on the operation of reactors, if reloads are not available at the right time and in the right place. Fuel managers and their company risk assessors have to manage their fabrication strategies carefully, to ensure that they have the right fuel available at the right time, without holding excessive stocks of expensive fuel bundles. There are different strategies to address this, giving different levels of security, but with fewer fabricators today, the cost pressure on fuel managers and the trend to minimise outage lengths, the risks of a disruption have undoubtedly increased.

NUCLEAR FUEL

Lost revenue for a large reactor would be of the order of $2 million per day, but the eventual cost may be even greater if the utility has to buy replacement power on the expensive open market. These considerations are prompting some utilities to move away from a pure 'just in time' philosophy for new fuel assemblies, towards keeping full reload batches of assemblies in inventory, irrespective of the carrying cost.

In summary, fuel fabrication is definitely a distinctive market and has rather different fundamental characteristics to those of uranium, conversion and enrichment. Despite the apparently comfortable capacity situation, at least in contrast to the other sectors, there is a lot of market activity that could change things. The full impact of the round of consolidation has yet to be seen, while prospects of new reactor orders are becoming more encouraging.

The uranium boom: was it predictable?

The world uranium market has suddenly become interesting once again, with prices (in US dollar terms at least) peaking at over $130 per pound in mid 2007, after the long period at around $10 per pound in the period from the mid 1980s to 2003. Prices started rising then and really took off in 2006 and 2007. After the peak, they fell back somewhat in the second half of 2007, but still ended the year at around $90, or nine times above the level of the 'uranium glut' of the '80s and '90s.

Talk of a uranium glut over that period looks somewhat odd when you consider the market fundamentals. For several years, non-specialists were told a story at general energy conferences, which many never seemed fully to believe. They were told that about 440 power reactors around the world consumed 65,000 tonnes of uranium each year, yet world production stood at only 35,000 tonnes or so. Pretty graphs were produced to demonstrate this point and, moreover, that a similar situation had existed since 1985. The explanation for the gap was, of course, a significant inventory draw-down, so huge that uranium prices were depressed. Indeed depressed to such a point that most of the mines went out of business, with only the lowest cost operators staying in production and able to make a small profit.

Cue looks of disbelief from the audience! People are familiar with markets being supplied by inventory from time to time but continuing for 20 years? With production only 55-60% of demand, surely prices should be strongly rising to stimulate new production to fill the gap? Sorry dear audience, I'm afraid not. Well, at least not until now.

The story could then progress to explain the mysteries of the inventory draw-down, in other words all of the so-called 'secondary supplies'. These add to primary production to ensure that supply equals demand in each year – which it clearly must as all the reactors have fuel and are generally running very well. Things get complicated here and audiences always seemed to suspect that presenters were doing this quite deliberately, in order to obscure matters. Or else they suspected that the basic figures were wrong, with demand overstated and production understated.

Leaving this aside, the colourful graphs could be broken down into the slightly different historic patterns in the West and the former Soviet Union and its satellites. Eras of uranium production for military applications could be discussed, with some of the material mined up to 50 years ago entering the civil market as a consequence of nuclear disarmament in the United States and Russia. Then of booming and declining civil reactor programmes, which led to significant inventory build-ups and run-downs. Variable tails assays at enrichment plants could be explained and the possibility of sending stocks of depleted uranium back through the plants to create more reactor-assay fuel. And stocks of reprocessed uranium and plutonium from civil reprocessing plants, which can be recycled in various ways as fresh fuel. Finally, the importance of trade in nuclear fuel could be mentioned, particularly between the East and West. Russia and its former satellites have effectively covered, since the late 1980s, a substantial part of the shortfall in European and North American markets.

The end of the story was always a plea for the impoverished uranium miners. Surely things could not continue this way forever? Eventually the secondary supplies must become progressively exhausted and then lots of new primary production would surely be needed. Maybe not so much that the gap closes completely, as it will take a long time to recycle in the civil market all the possible ex-military uranium and plutonium, while MOX fuel and reprocessed uranium may continue to cover a small segment of demand.

Yet until recently, not much happened to bring forward the prospect of new mines. Indeed, quite the opposite. Primary production became concentrated in fewer and fewer mines in a small number of countries, such as Canada and Australia, owned by robust companies that pushed operating costs down to the absolute limit. For most of these, thoughts of major new mines had to be put to one side as they struggled to keep existing operations going in adverse market conditions.

On the fuel buying side, many of the companies operating nuclear plants appeared to be seduced into a false sense of security. Industry analysts who presented supply and demand analyses over the years could perhaps be accused of 'crying wolf'. Estimates of when the secondary supplies would run out were continuously proved wrong, so the era of relatively easy supplies at low prices became prolonged – perhaps to the point that buyers believed that things would never change. Or at least not in their own working lifetimes.

The spur for the sudden change in 2003 was possibly the supply disruptions at some important mines, such as Olympic Dam in Australia and McArthur River in Canada, indicating that the primary supply infrastructure was relatively weak. A key factor, however, was a fundamental change in the way the Russians viewed their participation in nuclear fuel markets. Back in the late 1980s, when they first became a significant force in the market, they owned considerable excess inventories of uranium of various assays, from highly enriched uranium (HEU) right through to depleted uranium. In addition, they had surplus enrichment capacity in efficient centrifuge plants. Their main motivation was to cash in on this, as they needed as much hard currency as possible. Nuclear fuel was one of the few areas where Russia had any competitive advantage over the West and the demand for supplies from Europe and North America was assured (although sometimes constrained by various trade restrictions). So the Russian inventories were a significant factor in bridging the supply gap. This earned them significant revenue, but then their strategy undoubtedly changed.

Clearly their inventories were significantly depleted by the early years of the new century, so they began to regard what was left rather differently, as scarce and strategically valuable. Looking back, they may now believe that they sold too much to the West and also too cheaply. By supplying so freely, all they succeeded in doing was to depress the price for everyone for longer. In addition, the Russians have always regarded reactors they have built as their captive market. The numbers of these are increasing with VVER-1000 reactors completed in China and still under construction in India and Iran. Now the Russian domestic reactor programme is also showing signs of getting back on track. The Russians have also realised that their key strategic resource is their abundant oil and gas reserves. The West is developing an increasing dependence on these, like a drug addict seeking another fix. So better to hang on to the remaining nuclear fuel supplies and export the oil and gas. Their own uranium resources are limited and of relatively poor quality, so it would be better to channel most efforts into the production of other commodities.

The impact of this change has been felt throughout the uranium market over the past few years. There has been a sudden realisation that the Western supply infrastructure is very fragile and that primary production must be increased sharply and as soon as possible too. Yet uranium producers always said that it takes a significant time to get new mines up and running. The right economic incentives are essential but the regulatory burden today is very onerous.

The eventual spurt in uranium prices was therefore somewhat predictable but delayed so long that some people had almost given up on it ever happening. Because it was delayed, seemingly forever, the strength of the price rebound has undoubtedly been magnified.

Uranium prices: providing appropriate signals?

In terms of the currencies of most producing nations, the sharp uranium price increase since 2003 has been less significant, given the appreciation of their currencies against the US dollar. Yet the rise has become extreme and also sustained. Back in 2003 and 2004, those market observers who expected a similar pattern to 1996, when the price spike from $10 to $16 per pound was soon reversed, have been very disappointed. Some fuel buyers initially continued to sit on the sidelines expecting the price to eventually fall back, but soon saw that they were running the risk of being left out in a rush to secure adequate supplies for the next period. Taking a longer-term perspective, everybody had been sitting on the side of the court, waiting to see what happens, and there was always the risk of a scramble for supplies when everyone realised this. Everyone knew that procurement decisions made today would influence the amount of investment in new capacity to fuel future demand. The question for both buyers and sellers was always: Does the price today adequately reflect the future scarcity of supply? Each buyer and seller has to have his own answer to this and act accordingly.

It is clear that the buyers have a responsibility to help shape the future, but there are a large number of them who cannot necessarily be expected to act collectively in a perfect manner. Yet within the enrichment sector, it is notable that some have taken matters into their own hands by supporting Urenco's National Enrichment Facility plant in New Mexico. With uranium, on the other hand, it is only recently that major buyers, such as the Japanese, Koreans, Chinese and French, have invested directly in new mines in order to secure their own future supplies. Western buyers still seem to be maintaining faith in the development of an efficient uranium market to accomplish this, but this may require extreme price volatility, which could prove unsettling.

Everybody agrees that more uranium supply is needed, particularly for the period beyond 2010 when world utilities are less well-covered by current contracts. Supply and demand will of course adjust to price, but both are rather inelastic. Once reactors are up and running, they need to be fuelled and their operators must pay the going rate for supplies. Their suppliers are essentially competing to meet this essentially fixed (in the short term) level of demand but cannot easily adjust volumes either. Hence there is always a risk of price volatility when mismatches suddenly become apparent. Yet it is difficult to get uranium out of the ground. There are a myriad of problems such as obtaining finance and regulatory approval, trade restrictions and other government issues, technical glitches at mines and transportation difficulties.

Some of the supply, indeed, is not so sensitive to the uranium price. A substantial part (10-15%) of uranium supply has historically been produced as a by-product of other metals (usually copper and gold) with the economics depending also on the market position of the other product(s). Then there is the secondary uranium originating from nuclear weapons, which is arriving on the market mainly for non-proliferation rather than economic reasons. Moreover, uranium supply which comes from re-enriched depleted uranium ('tails' material) will diminish in quantity at higher uranium prices, as customers will choose a lower tails assay at enrichment plants and there will also be less spare enrichment capacity available to then re-enrich the tails.

Given enough time, the market should somehow work out. Without foolish government intervention and trade restrictions, there is no reason why prices should not get fixed at levels which guarantee sufficient production. But the combination of demand inelasticity and temporary (at least) supply inelasticity could lead to prices, at least in the short term, being very volatile. Indeed, it is still rather early to judge whether the price increases since 2003 are sufficient to bring on enhanced supply on a timely basis. The magnitude of the price increase suggests it should do so, but we're in unchartered territory here. Hundreds of companies are now showing greater interest in getting involved in uranium production but this interest must eventually get translated into pounds from the ground.

One possible consequence of prices standing at high levels for a prolonged period of time is that it could potentially bring on some relatively high cost mines, which could overhang the market if prices begin to retrench. Once mines start up, capital costs are effectively 'sunk' and they will continue in operation so long as they can cover marginal operating costs. Overproduction and a renewed period of low prices is thus a risk, particularly if new reactor build turns out lower than expected. The uranium market is certainly competitive, but the number of companies involved has been driven down sharply in the previous market conditions. The survivors are well aware of the nature of the competition but will probably not want to see a lot of new companies arriving on the scene, solely chasing higher prices which could prove temporary. Their reaction may be to acquire the most promising 'new' producers at an opportune time. In any case, securing long-term contracts at the higher spot market prices will tap much of the potential demand such companies would be targeting.

Larger producers such as Cameco and Areva certainly have sufficient market power to exert some influence over the market. Both are involved in managing the introduction of secondary supplies into the market, such as the uranium component of the Russian HEU. They can also buy uranium on the spot market whenever the price is weak and sell when the price is good to try to remove some of the market volatility. Yet they still need sufficient prices to make a reasonable profit from the new mines such as Cigar Lake and Midwest, which may have remained on hold had prices not risen. They therefore have to play a delicate balancing act within the underlying market fundamentals, which they don't have the power to affect in any significant way.

Indeed, the spot uranium market is diminishing in importance as the longer-term contract market reasserts itself as customers wish to secure their long-term positions. Most market participants will welcome this as the tail has been wagging the dog for too long with the market awash with secondary supplies which threatened to prevent the creation of necessary long-term price signals. Utilities could always rely on picking up supplies on the spot market at competitive prices without giving the suppliers the level of return needed for them to invest for the long term. This has fortunately now changed and there are some expectations that the uranium market may eventually become more like other commodity markets, with a good balance of longer-term and spot dealings.

Nuclear power is a long-term business, so one would naturally expect there to be a large number of long-term contracts for fuel supplies which the producers can use as collateral in borrowing money to invest in new mines. Indeed, this is how things used to be 20-30 years ago, when many new mines were developed to fuel rapidly-growing nuclear power programmes. The period since then may eventually be seen as rather an aberration, caused by excessive

Uranium market:
what needs to happen?

secondary supplies, with sub-$10 prices a distant and, for the utilities, fond memory.

Quoted uranium prices continue at historically high levels. After being stuck at around $10 per pound for many years, prices rose to over $130 in mid 2007 but have now stabilised somewhat at around $70 (April 2008).

The extent of the price rise has created a number of issues for the industry with anti-nuclear opponents, for example, claiming that the price spike indicates that there won't be enough uranium to satisfy future reactor requirements, particularly if there is an upsurge in new reactor building. There has also been a view expressed that such price rises may also begin, eventually, to have an impact on the economics of nuclear power relative to other electricity generating modes. The first of these can easily be shown to be false. There is plenty of uranium in the ground and it is highly unlikely that rectors will ever go without fuel. In any case, if reactors were to run out of fuel, surely that is an issue for investors and not nuclear opponents – one would expect them to be extremely happy with such a prospect! The economics issue is more complex. For nuclear utilities who are price takers in competitive electricity markets (unable to pass on cost increases), every extra cent paid out on nuclear fuel impacts their bottom line profit. But it is highly unlikely to stop them operating their nuclear plants – just cut already good profits. For new reactors, the fuel element in total costs is even less significant, representing only about 15% of the total. It is more the talk of uncertainties in the nuclear fuel market and, in particular, worries about the fragility of the supply infrastructure, which could impact on investment decisions. Investors require comfort on this – the list of possible risks in any nuclear plant investment is already long enough.

There is, however, another issue surrounding the price rise that is worthy of some discussion. This concerns the mechanics and price reporting of the uranium market itself. It can be argued that the long period of price depression followed by such a dramatic spike indicates that the uranium market is not functioning as it really should. The explanation for the way prices have behaved is ultimately quite simple. They were depressed for many years by abundant secondary supplies, which pushed them below the production costs of many mines, which then had to close. When it became clear that secondary supplies were beginning to become exhausted and there were signs of new reactor orders, an upsurge of market demand met supply which was effectively fixed in the short and medium term by years of low investment. Hence prices started to rise. Additional elements of demand were triggered by this, including a speculative element from financial companies, thus serving to push the price up further. An efficient market, however, should transmit information about supply and demand in such a way that prices move smoothly in such a way to stimulate the required production in a timely manner, so that future demands can be comfortably satisfied. Yet we know commodity markets tend to be volatile, with prices tending to overshoot at the top of the cycle and then remain depressed for rather longer than is fully justified by the future production needs. Uranium may just be an extreme case of this, but with a longer cycle length.

Nevertheless, the particular characteristics of the uranium business may ensure that it should arguably have much calmer market conditions. Electricity companies have to go through a prolonged planning and licensing period before a reactor goes into operation, then they aim to run it intensively for upwards of 40 years into the future. Uranium mining companies have a very similar time horizon – again it can take many years to bring a new discovery into production, but once this happens, the mine will run for many years. So with similar time horizons, the two sides, buyer and seller, ought to be able to easily do business together. In the past, arrangements such as buyers taking equity stakes in uranium mines or granting long-term supply contracts at favourable rates (thus providing mining companies with suitable collateral to borrow money for mine investment) were very practical solutions to address the common interest in secure and orderly supply at reasonable prices. The long period of depressed prices, however, left the buyers feeling that such arrangements were against their interests. As it was possible to buy as much uranium as they needed at very favourable prices, why take the risk of investing in mines or grant long-term contracts, where the prices could well be significantly in excess of the spot price?

Now the boot, formerly very much on the foot of the buyer, has dramatically switched to the seller. Again this is typical of all commodity markets, but this has happened so swiftly that there is little time to consider whether the arrangements in the market are sensible or should be changed to the ultimate advantage of all. In the earlier period, the buyers didn't want to change anything, but now it's the sellers who are perfectly happy. There has been some movement back towards buyers taking equity stakes in producing mines (notably the East Asian utilities and their agents in Kazakhstan) but few signs that buyers are talking to mining companies about future production costs from their mines and negotiating long-term contracts based on these. Indeed, any mention of cost of production in the current market environment tends to be laughed at. Both short- and long-term supply contracts are apparently being fixed on the basis of the spot price at the time of delivery (which effectively brings together the marginal quantities of material available for short-term delivery). This makes little sense for a commodity where both supply and demand have such a long-term focus. It makes it impossible for utilities to budget future fuel costs and means producers may once again eventually receive very low prices for material, if prices plummet and any floor provisions in their contracts run out of time. This may make producers wary of carrying out the high level of capital investment now required to bring supply and demand into good balance for the future.

Much of the problem relates to lack of market liquidity and transparency. Prices are published on weekly and monthly bases by informed observers such as Ux Consulting and TradeTech and are based on information they glean from market participants. They would be the first to support receiving increased knowledge about many more transactions, but it doesn't currently exist, as visible deals are few and far between. More recently, auctions of small quantities of available material have provided much of the information. There are then 'transactions-derived' price reporting systems, such as UPIS and Uranium Online. Both of these suffer from significant weaknesses (in the case of the former, limited coverage and time lags and in the latter, very few transactions to date). Further information can be gleaned from company announcements and accounts, where average selling prices may be quoted or can be calculated. More recently, uranium contracts have been quoted on the NYMEX market, but volumes have so far been comparatively small. Finally, there are historic price series from governmental bodies such as the Energy Information Administration in the United States and Euratom in Europe, which give an authoritative rearview mirror but little else.

How can things change for the better? Rather like the somewhat dated infrastructure of much of the nuclear fuel supply business, the current uranium pricing system is a prisoner of its past. The infrastructure issue is slowly but surely getting addressed by the required new investment, but the uranium market is still in need of something new. The first requirement would seem to be a spot market of greater transparency and with much more volume. Of course, everyone supports more transparency, so long as it's other people providing it, but market participants should gradually be willing to support it. The financial investors who have entered the uranium market are potentially important players here. It is unclear whether the market can become more akin to the other energy markets like oil, gas and coal, where there is a greater range of financial exchange trading rather than physical trading. The size of the market will always be considerably smaller in uranium but the interest expressed by many 'outsiders' suggest that it could potentially be sufficient to promote an active market. This will require standard contract sizes and delivery terms, online screen trading with small lot sizes and also plenty of free inventory to lubricate the market.

None of these requirements are yet much in evidence, but could surface over a period of 3-5 years. In particular, the uranium buyers would need to be more proactive in their procurement decisions, buying smaller lots to give them a better-balanced contract portfolio, rather than occasionally entering the market to buy large quantities. This will require big cultural changes within their companies, breaking with a lot of past practice. Then the producers need to start producing much more uranium to free up quantities available on the spot market. This will possibly be associated with a fall in the spot price from the peak (wherever this is established) but will be the ideal time to spur new market arrangements. The point where the interests of both buyers and sellers coincide may be a very short one, but is the key to major change.

The question then remains about the longer-term market. To some extent, having a more transparent and liquid spot market will go some way towards addressing this. Assuming an improved spot market has a futures element, ideally with a forward price curve showing future bids and offers, there may be much less market volatility, with some of the peaks and troughs rounded off. It would then be more realistic to index prices for delivery in the 3-10 years timeframe to these prices, perhaps incorporating other indices, including general inflation. This is up for negotiation by individual buyers and sellers who, as always, will develop their own specific contract types. But there must still be some room in the market for standardised longer-term contracts, with fixed quantity and delivery obligations. Perhaps these can be considered as a very worthwhile second stage after the short-term market is reformed. A credible uranium market-maker may get established in the spot market, with the ability to take things further. One possible danger with long-term contracts is that either party may conceivably try to get out of the arrangements if the spot price moves too far from the contract price.

In conclusion, if nuclear is to take its rightful role as a major feature of the 21st Century energy scene, it is important that there is as much clarity as possible about all aspects of its uranium raw material base. The uranium market still retains arrangements developed in the mists of time, but now is an appropriate time for something new to be developed. It will undoubtedly take some entrepreneurial leadership to induce this to happen, plus realisation that the current marketplace cannot be in the long-term interest of either buyers or sellers.

Nuclear fuel: some positive changes?

After years of being rather stuck in a time warp, it now seems there are some potentially significant changes taking place in the supply of nuclear fuel. The sharp rise in the uranium price has been much analysed and discussed, along with the arrival of hundreds of new junior uranium companies, a much enhanced exploration effort and the prospect of a significantly greater level of uranium production by 2010. The considerable investments in new centrifuge enrichment capacity have also been heavily covered in market commentaries. There are, however, some other aspects very worthy of note.

Outsiders, such as financial analysts and consulting companies, always find extremely curious the way in which nuclear fuel buyers buy uranium, then have the fuel converted, enriched and fabricated by separate service companies before being loaded into a reactor. "Why don't they just buy the fabricated fuel?" they all ask. This is, of course, a good question. The answer is a mixture of history, the bad experience of the famous Westinghouse case of the late 1970s (when the company was contracted to sell fabricated fuel at prices below costs that were inflated by unforeseen rapid uranium price increases), some inertia or conservatism in utility fuel managers (some cynics would see it as enlightened self interest in order to preserve their jobs) or, more positively, a view that the present arrangements guarantee the best prices for the buyers (which they indeed may). Finally, the fuel fabrication part of the fuel cycle is different to the uranium, conversion and enrichment sections – it is not a fungible product or service but a high technology specialised operation, with products specific to particular reactor designs (or even particular buyer requirements).

The usual procedure may, however, be under renewed attack. Some of this has possibly been prompted by the extent to which many fuel buyers have been burned by the uranium price increases. For some US utilities, which back in 2003 had little inventory and low contract coverage for the 2005-2010 period, the pain has been considerable. Why not take up a decent offer for the supply of a total fuel package at a predictable (and hopefully currently competitive) price?

The problem up to now has been the lack of willing sellers.

Buying enriched uranium product (EUP), *i.e.* uranium in its enriched form, is a halfway house and enrichment companies have been happy to oblige. They are effectively in the uranium and conversion business themselves, with their ability to under- or overfeed their plants and re-enrich depleted uranium. But now we are seeing vertical integration in the front end fuel cycle, with other companies hoping to copy Areva in its ability to offer all four components in a package. So far this has been only for first cores and limited refuelling for new reactors, but may eventually go much further. The Russians have, of course, for many years provided a full fuel service for the reactors they have sold overseas (including taking back the used fuel). Now Kazatomprom, the rapidly-growing uranium producer from Kazakhstan, has shown by taking a 10% stake in Westinghouse (now majority-owned by Toshiba) its eventual intention to offer a complete fuel service, by signing a cooperation deal on a conversion facility with Cameco. Kazatomprom already has a large fuel pellet producing plant and seems set on extending its fuel fabrication activities much further – it wants to add further value to its significant natural uranium endowment. Other fuel cycle participants are likely to follow this trend.

Indeed, offering a full fuel service also fits in with latest international moves to provide assurance of fuel supply to new countries building nuclear power plants, in return for them renouncing their right to build domestic enrichment and reprocessing facilities. This could still involve them buying each component separately, but does seem to lead itself to them buying not only a reactor, but many years of assured total fuel supply at the same time.

Another interesting trend has been some signs that nuclear utilities are prepared to invest directly in fuel cycle facilities to secure future supplies, also to stimulate infrastructure renewal and spur competition. This was initially seen with the investments in Urenco's New Mexico National Enrichment Facility by US utilities. These have been more reluctant, however, to invest in new uranium mines, in contrast to the position in the past when many took stakes in uranium companies. This often ended in tears, however, with many years of poor uranium prices, and there is clearly a reluctance today to risk getting burned once again in businesses very different to their own. It has mainly been the Japanese, Korean and Chinese utilities that have been investing in mines, particularly in Kazakhstan, but also in Africa and Russia. The tight short-term uranium supply situation has clearly motivated this, along with a desire to free themselves, at least to some extent, from the vicissitudes of the uranium market. With most near-term nuclear growth currently set to be concentrated in the Asian region, what is seen as the world uranium market today may increasingly become irrelevant. Companies may take their equity shares of production at cost prices, rather than based on references to quoted uranium prices. Unless there is substantial new nuclear build in North America and the European Union, the uranium market of today may become merely a declining residual feature, with somewhat quaint and mysterious practices largely divorced from the worldwide industry trends.

Seeking ways to get as far away as possible from uranium market gyrations is clearly a popular theme. It is clear that the market, as it stands today, doesn't work to anyone's long-term interests. Having swung from a position for over 20 years when prices were too low to offer a workable financial incentive for badly-needed new production facilities, the opposite could now be happening. Fixing contract prices for a vital commodity in a very long-term business like nuclear power, based on short-term surpluses or shortages in an illiquid, opaque spot market is clearly ridiculous in any economic sense. This practice led to the situation of weak supply infrastructure today and the consequent need to ramp up production very rapidly. Yet the seeds of an unpleasant subsequent bust are already sown. Establishing new production facilities may take longer than generally expected but come online they undoubtedly will – the economic advantage of doing so, based on current spot prices and likely cost levels, is just too obvious. The risk, however, is that prices will fall too early and too far and fast, potentially choking off some of the expected production growth, as company share prices and their ability to raise money take a hit. Although nobody would begrudge today's uranium companies enjoying the much higher prices after being financially hammered by buyers for so long, they must be careful not to 'milk' the situation as the buyers foolishly did (only to get caught out now). Far better to return, one would think, to signing long-term contracts with buyers that give the producer an adequate return on investment and the buyer some assurance about future price levels as well as on firm deliveries. Indexing future prices to spot prices at the time of delivery seems a distinctly odd practice, when a standardised but tradable long-term contract market would be a much superior option.

Finally, it now appears from recent announcements and US court rulings that we may be inching forward, at last, towards a complete world market in nuclear fuel. The main distortions at present are that Russia makes it difficult for its 'captive' customers (with

Russian-designed reactors) to take fuel supplies from elsewhere (mainly by offering very good prices), while the US market in enrichment has, in turn, been protected from Russian primary supply (although downblended Russian HEU provides around half of the US market). Areva and Urenco from Europe have also faced trade actions in the US enrichment market. All of these seem to be slowly withering away but a fully competitive world market is still a long way off. Ukraine seems set to load Western-origin fuel in some of its reactors, to follow the Czech Republic, which did so for a time. Russia will carry on making hard-to-beat offers but East European reactor operators will continue to look Westwards after joining the European Union (EU). US utilities will gradually be able to sign contracts for Russian enrichment as the HEU deal expires in 2013 and it seems only a matter of time before the US market is opened up completely. Restrictions imposed by Euratom in the EU seem gradually to be withering away too. Utilities should surely now be treated as 'big boys', quite capable of looking after their own interests in supply security and prices. Nuclear will hopefully gradually become treated as just another business rather than something requiring a lot of market interference.

In conclusion, a number of areas are showing very positive developments for nuclear fuel, such that it will be ready to play its important role in a future and prosperous nuclear industry. One cloud on the horizon, nevertheless, is the continued transportation difficulties the industry faces, which threaten to imperil the economic logic prompting sourcing material from the lowest cost locations, subject to security of supply considerations. These require maximum attention to get resolved or else all the additional options the buyers can now begin to consider may become meaningless. But overall, it is possible to foresee the fuel supply infrastructure getting renewed at long last, together with a heightened range of market alternatives put under offer.

5. TRADE AND THE BACK END OF THE FUEL CYCLE

Introduction The nature of nuclear technology and the materials used by it has always meant that there have been significant international restrictions on knowledge transfer and trade. These are bound up very much within the provisions of the Treaty on the Non-Proliferation of Nuclear Weapons (NPT) and its policing by the Nuclear Suppliers Group (NSG). Restrictions on countries which fall foul of the rules, such as India and Pakistan, are stringent and have served to constrain their civil nuclear development. Beyond this, restrictions on trade in nuclear fuel and technology are not particularly stringent, particularly if nuclear cooperation agreements are in place between bilateral partner companies. There is a high degree of specialisation amongst suppliers within the industry today and trade is a necessary adjunct to this. Today it is the difficulties of transporting nuclear materials that are giving cause for most concern, threatening to bring severe disruption to the most efficient development of nuclear commerce.

What to do with the used fuel when it is discharged from a nuclear reactor has been one of the biggest issues since the early days of nuclear power and has given rise to probably the biggest issues on public acceptance. It can now be seen that the industry scored an 'own goal' by easily accepting that this material should be regarded as waste, before it became clear whether or not it could have economic value in the future. The use of language is very important and as soon as something is categorised as 'waste', a solution must be found for dealing with it very quickly, as it cannot be passed on as a liability to future generations. What to do with nuclear high-level wastes has been much-discussed, with international scientific consensus now very much in favour of deep geological repositories in favourable locations. Even if used fuel is reprocessed, there will eventually need to be repositories, although the size of these will be much reduced. There is now some movement back towards the idea of reprocessing (in other words the 'closed' nuclear fuel cycle rather than the 'once through' cycle), and also increased acceptance of the idea of long-term surface storage before this.

Reprocessing itself has always been a topic of significant controversy. Although it has been advocated by the industry as promoting sustainability of resources and also a sound overall waste strategy, it has been demonised by anti-nuclear forces because of the separation of plutonium. Although plutonium is certainly not such a dangerous material as often claimed, the increased focus on the proliferation risk means that future reprocessing plans must incorporate a technology shift beyond the current Purex process. These plans look many years ahead and, in the meantime, some countries will continue to develop their closed fuel cycle strategies using currently-proven technologies, notably Japan. Increased world uranium prices suggest that recycling uranium and plutonium should now have increased financial advantages.

As nuclear becomes a more mature business, decommissioning of facilities naturally becomes a relatively more important activity. Particularly in the United Kingdom, which had an early start in nuclear and engaged in a lot of experimentation of various technologies back in the 1950s and 1960s, cleaning up old sites has become, financially at least, the major focus in the industry today. To some, this smacks of an industry in irreversible decline, rather than a natural, necessary stage in an industry, which should still have a bright future. The huge sums of money to be expended that are mentioned will in fact be spent over a long period of time. The decommissioning expense is actually not that material in the economics of new nuclear reactors and the provisions made may well eventually prove to be too generous, as costs of decommissioning should fall as more experience is gained.

Nuclear commerce: do restrictions hinder future growth?

The highly technical nature of nuclear power ensures that there are specialised knowledge and materials that countries must acquire before they can have a reactor construction programme. Under the Treaty on the Non-Proliferation of Nuclear Weapons (NPT), the five nuclear weapons states are obliged to transfer technology for the peaceful uses of nuclear science to treaty signatories who have given up any interest in weapons. It is sometimes argued that the weapons states haven't fully fulfilled this, but the spread of (initially) research reactors around the world and (more recently) power reactors to over 30 countries, suggests some success at making nuclear something like a 'normal' business, where new technology developed in one country rapidly spreads elsewhere. There is also acute specialisation within the nuclear fuel cycle, which requires a trading regime and provision of transport. Buyers seek supplies at the lowest cost, subject to concerns on security of supply, and uranium supply and the other front end nuclear fuel activities are now largely in the hands of a limited number of companies. To the extent that there are impediments to the free flow of technology and materials today, how significant are they and could they constitute a barrier to the future development of the nuclear industry?

The Nuclear Suppliers Group (NSG) polices nuclear commerce from the point of view of adherence to the non-proliferation regime, through the implementation of Guidelines for nuclear and nuclear-related exports. The NSG Guidelines are followed by 44 national governments in accordance with their own laws and practices. Decisions on export applications are taken at the national level in accordance with national export licensing requirements.

The NSG was created, initially with only seven members, after India first exploded a nuclear device in 1974, indicating that nuclear technology supplied for peaceful purposes could potentially be misused. The Guidelines are essentially a set of export rules, seeking to ensure that transfers of nuclear material or equipment are not diverted to unsafeguarded nuclear fuel cycle facilities or towards nuclear explosive activities. They also recognise the need for physical protection measures in the transfer of sensitive facilities, technology and weapons-usable materials. The first set of NSG Guidelines governed the export of items that are especially designed or prepared for nuclear use, such as fuel and reactor equipment. These were extended in 1992, by the addition of Guidelines for transfers of nuclear-related dual-use equipment, material and technology, essentially items that have additional substantial non-nuclear uses, for example in industry.

The main practical impact of the NSG is that nuclear trade with India and Pakistan has been severely limited, as neither country has signed the NPT while developing nuclear weapons. This has meant that each has had to develop its domestic nuclear sector without recourse to much assistance from outside. India, in particular, has very ambitious nuclear plans, but suffers from a shortage of good uranium resources. It has based its programme on the use of as much recycled fuel as possible and notably on fast reactors and the subsequent development of full-scale thorium-based reactors, as it has significant thorium reserves.

Neither India nor Pakistan are ever likely to sign the NPT (on the basis that the treaty discriminates against them as second class nations and refuses what they see as their legitimate right to possess nuclear weapons for defensive purposes). Other ways are being sought to end their isolation by wider diplomatic efforts and the proposed agreement between India and the United States is in this vein. Difficult issues surrounding other countries that have signed the NPT, but are accused of ignoring its provisions, have occupied a lot of media attention in recent years. Iran and North Korea have both faced economic sanctions for their alleged activities.

Very few countries possess the full range of facilities required to carry out all steps of the nuclear fuel cycle. The degree of specialisation in the nuclear fuel industry clearly contributes to the overall economic efficiency of the nuclear fuel markets, as it would be prohibitively expensive for a country with a small or fledgling nuclear power programme to develop all the necessary fuel cycle facilities. International trade in uranium and nuclear fuel cycle services is a necessity for the production of nuclear electricity in most of the countries presently using nuclear power. The need for non-proliferation controls means that governments have always paid careful attention to the uranium and nuclear fuel markets.

The social, political and economic importance of the security of electricity supplies means that assurance of fuel supply is of overriding importance for utilities. Diversity of supply sources is often the main way that utilities ensure they will receive sufficient nuclear fuel to meet their requirements, particularly in countries with no domestic production. There are now plans to offer supply assurance for any new nuclear country renouncing the right to build domestic proliferation-sensitive enrichment or used fuel reprocessing facilities.

When the General Agreement on Tariffs and Trade (GATT), the main treaty governing international trade, was established in 1948, trade in fissionable and related materials was excluded from its provisions on security grounds. Nuclear trade is also outside the scope of the World Trade Organization (WTO). Thus there is no multilateral framework for the conciliation of disputes, such as the 'dumping' cases raised in 1991 in the USA against uranium producing countries from the former Soviet Union. Such disputes must therefore be resolved through bilateral negotiations and agreements between the exporting and importing countries.

The main direct government barriers to trade in uranium and nuclear fuel services are the import tariffs on fabricated fuel imposed by many countries, the uranium export monitoring policies of Australia and Canada, and the uranium import policies of the United States and the European Union (particularly in relation to imports from the former Soviet Union). These have all been relaxed somewhat in recent years but have still have some impact on the market.

In addition, Russia is very protective of the fuel supply contracts for any reactors it sells outside its own territory, effectively leading to a continued segmentation of the market by reactor type. The main impact of the US restrictions is now to prevent local utilities from contracting directly with Russian suppliers for enrichment services. This is facing increasing pressure from both sides, who would very much like to do business with each other, and is therefore likely to ease over time.

Concerns about proliferation of nuclear weapons and the imposition of some trade restrictions to protect domestic producers have led to a web of rules and regulations concerning the movement of nuclear materials. The basis for these rules is the concept of 'origin' of the material. This is common throughout all world trade (largely because of the imposition of tariffs and anti-dumping rules) but is particularly important owing to the proliferation concerns in nuclear. The first origin assigned to a nuclear material is the mining origin of the original uranium but it acquires other customs origins at each stage of the fuel cycle *i.e.* conversion, enrichment and fuel fabrication. The concept of 'substantial transformation' is important, as this is where the customs origin changes. However, this term is defined differently in various countries, so there is no harmonised and comprehensive international set of rules on origin. Nuclear materials also have 'obligations' attached to them, which are rules assumed by importing governments in accordance with the requirements of the governments of supplying countries. Typical obligations are those imposed by uranium exporting countries on the use of the material. For example, Australia has only exported uranium to France on the basis that it will not be used for military purposes.

There is also substantial swapping or exchanging of nuclear materials. As materials are fungible (in other words, easily substituted by something physically its equivalent) this is, in theory, quite easy but in practice is a hugely complex matter. Obligations and sometimes the origin of nuclear material can be swapped and this accomplishes various goals, chiefly the reduction of transportation cost and risk.

Indeed, transportation has become probably the biggest concern within nuclear commerce. Given the degree of specialisation that takes place in the nuclear fuel cycle, there is accordingly the need to move materials around the world, with any interruptions to this likely to have substantial consequences, financial and other. All procedures employed are primarily designed to ensure the protection of the public and the environment. Nevertheless, most transports of radioactive materials are not nuclear fuel cycle related, as they are used extensively in medicine, agriculture, research, manufacturing, non-destructive testing and minerals exploration.

The international shipment of radioactive material – known as Class 7 under the UN's dangerous goods code – is becoming increasingly difficult, with fewer shipping companies and ports willing to accept this cargo. Marine routes present the greatest challenge, with long diversions commonplace, notably in relation to shipments of uranium and other front end nuclear fuel materials and higher-activity radioisotopes. In South America and southern Africa, suppliers sometimes have to transit another country to find a port willing to accept their cargo. From the standpoint of the carriers and ports, Class 7 clearly represents disproportionate difficulty compared to the small amount of business it provides. Because of the complex procedures surrounding Class 7 and differences in interpretation of international regulations, vessels have been held up, incurring significant costs. In addition, as nuclear cargo needs to be shipped according to a route approved by the competent authority, it reduces the flexibility to change routes and modes, as usually built into transport operations.

In summary, it can be said that provided that nations fit in with the obligations imposed by the NPT, international nuclear commerce does not face insurmountable barriers. Indeed, by comparison with the trade in agricultural commodities, it can be argued that the rules and regulations in force today are not particularly onerous and should not prevent new countries acquiring power reactors if they wish to do so. Establishing enrichment and used fuel reprocessing facilities in such countries usually makes little economic sense, while offering them fuel assurance via international intergovernmental arrangements caters for the security of supply risk. With the general easing of governmental restrictions on nuclear material flows, it is concerns about transport that are now threatening the future of nuclear commerce. At the very least, this could lead to substantial cost increases, but could also threaten security of supply. Transport concerns are being addressed by establishing a better dialogue between government, the industry and the contractors themselves. Both port and carrier shipments need to be freed up in order to provide the confidence that is needed for a sound industry future.

Nuclear's Achilles' heel:
or own goal?

Nuclear is certainly a very long-term business. A reactor planned today may be online within 5-10 years, then will run for 40 or maybe even 60 years. Just like a long distance runner, not a sprinter. Using this analogy, the profusion of new gas-fired generating plants are the sprinters, offering quick returns to investors for a ten-second dash. Our distance specialist requires a higher investment in terms of his preparation and conditioning, but will then possibly produce a superior long-run return, blitzing the opposition. But unfortunately, our marathon man suffers from many injuries, the worst of which is his Achilles' heel. He overcomes the other pulls and strains, yet still often runs in pain owing to this suspect tendon.

Diligent nuclear advocates stand up in front of sceptical audiences around the world. The economics demon is slayed by colourful charts showing rising capacity factors, low fuel costs and high operating profits. The proliferation threat is disembowelled by reference to the International Atomic Energy Agency (IAEA), its good work and dedicated international staff. On safety, they quote the industry's tremendous safety record and then emphasise what a nuclear plant can do for an impoverished Chinese or Indian coastal town in terms of economic multiplier effects (well, it beats fishing!). It feels like the audience could be won over by eloquence matched by careful use of statistics, when a girl stands up, clearly with a good knowledge (and understanding) of the industry websites. "This is all very well," she says, "but what about the waste?"

"What waste?" thinks the industry spokesperson. Waste of energy in the world today, waste of water through cooling nuclear plants or waste of time me being here tonight? Unfortunately none of these; it's immediately clear she's talking about used nuclear fuel. And when the industry side fails to knock down her confident argument, she opines, "this is your Achilles' heel," to tacit nods of approval from most of those present.

Then the industry goes along with this assessment. "Yes, it's our Achilles' heel," we agree amongst ourselves, "we must really do something to demonstrate that our industry has workable, economic and environmentally-acceptable solutions to this problem." Indeed, we rather favour the language of the Achilles' weakness. Didn't the industry train very hard, doing the best it could to win, but regrettably suffers from an unfortunate vulnerability for no fault of its own? So it can't race any more and deserves public sympathy with good wishes for its unfortunate plight.

Yet isn't this complacent nonsense? Far from being the Achilles' heel of the industry, this issue surely counts as its most spectacular own goal, scored by an industry prone to such defensive lapses. Yet this one takes top prize! It's 0-0 with not long to go. The bulky central defender, taking a wild hack at an innocuous cross going nowhere, only succeeds in blasting the ball into the bulging net, through the stranded legs of his bemused goalie. Then destined to be replayed time and time again on TV, accompanied by the guffaws of the studio analysts. Cue chants of "1-0 to the Greens", from the away fans, the above lady amongst them, accompanied with bye-bye waves to the nuclear team's fans as they depart the stadium in their droves to their parked cars and loved ones.

Let's elucidate. In fact, the industry's team was earlier largely on the attack against the struggling Greens. It had possession of used nuclear fuel near their goal, but their forwards' lack of ball control let them down. It didn't even need a brave sliding tackle; somehow the spent fuel got immediately re-categorised as waste. Possession went to the Greens who thereafter scarcely gave it back, culminating in the spectacular farce at the other end.

The problem for the industry's side was quite simple. Once it became re-categorised as waste, the opposition began to make mincemeat out of the industry's spokespersons and the match was effectively over. If it's waste, something has to be done about it and quickly too. After all, it's a huge liability, dangerous too, not an asset.

No longer destined to be stored and then possibly recycled as fresh reactor fuel, it becomes a big problem. It cannot possibly be handed onto the next generation to deal with. But then the industry contradicts itself by saying, "well actually, the volume is quite small, no more than the magnitude of the football pitch itself, so don't worry too much, it's not a problem at all." Well is it or isn't it? Then arguments are adduced that the industry has workable solutions to the problem but just needs the time and money to demonstrate these 'final disposal' repositories. But then it's uncertain whether the material may have yet some economic value in the future, so builds in the concept of retrievability. But what's the point spending billions of dollars putting this stuff down a hole in the ground when you're maybe going to bring it back to the surface again? Maybe just before further re-categorisation takes place – no longer a repository, it becomes potentially a deep plutonium mine (as radioactive decay gets underway).

What a dog's dinner of confusion! It's not surprising that the industry's team could never get the ball back again and a rearguard action is mounted to defend the industry's goal. The Greens are rampant, all over the pitch.

What should the industry have done differently and how can it possibly rescue its bad position on this issue today. The first point is much more care on the use of language. Never, ever admit what you have is 'waste' until you are absolutely certain that it does not have (and never will have) any economic value. Rather like the fossil fuel guys denying that they're leading the world towards environmental meltdown through carbon-induced global warming. Don't admit it unless you really have to – make the opposition fight for their victory. Perhaps this is morally not a completely defensible position, but it makes sound business sense. Indeed, the worst example (which has ensured continued high profits) is the tobacco companies refusing to accept that their product is linked with lung cancer. But nuclear is certainly not as bad as that!

The second point is that the public find the once through cycle with deep repositories difficult to accept. If the stuff has to be put so far below ground, it must be really nasty and therefore is not something we want to pass onto the next generations. It is not really final disposal – the problem essentially remains. Hence the long-running saga over Yucca Mountain in the United States. The industry should certainly be demonstrating that such repositories are technically feasible, but it is debatable that they are needed now. Long-term storage on the surface, until the question of whether the spent fuel is an asset or liability, is a far better option.

Yet in the early days of nuclear power, back in the 1960s and 70s, it was believed that uranium scarcity would necessitate reprocessing of spent fuel and recycling in whatever form, eventually in fast reactors. This doesn't avoid the need for repositories but because there is a coherent spent fuel management policy, the industry's position on public acceptance is far better. As it has turned out, the combination of rapid uranium resource development and the slowing reactor programmes has meant that only a few countries, notably France, Japan and Russia, have remained attached to the recycling model (yet with fast reactors now in the even more distant future). This may have looked foolish from the economic standpoint when uranium prices were low, but now looks more sensible now the market has tightened. Inventories of reprocessed uranium which looked like a liability at $10 per pound become significant assets at $90.

Recycling spent fuel also fits in with the concept of sustainable development, in that the maximum value is extracted from the mined material. It is true that the current Purex reprocessing technology is less than satisfactory in an environmental sense as it produces significant quantities of lower-level wastes. But the policy of not rushing into repositories, storing spent fuel long-term and probably reprocessing it at some stage in the future looks the right one from all standpoints. The adduced greater proliferation risk has been brought up as a downside, but is really nonsensical. Terrorists or rogue states can do little with materials from these stages of the fuel cycle.

So what can the industry do in the future to get out of this mess? Probably four things. Number one, don't be afraid to be uncertain whether used fuel will be an asset or liability, as you can't be certain what future nuclear fuel markets will look like or how technology will shift. Try to sell the idea of long-term surface storage to the public on the basis that you are passing a potential asset onto the next generation, not a certain liability. Hence, 'used fuel' is better terminology than the often used 'spent fuel' as it implies that there may be residual economic value. Secondly, continue to investigate and demonstrate the technical merit of deep repositories as, whatever occurs, some of these are going to be needed in the future. Thirdly, look positively at the concept of international repositories. There are significant regulatory (and maybe public acceptance) problems with these, but the idea of each nuclear country having its own looks ludicrous from several angles. Finally, actively pursue research in improved technology in reprocessing, which should take place in a limited number of safeguarded sites around the world (as has also been suggested for enrichment facilities). The world could well be short of nuclear fuel in the coming decades, as was originally predicted, so this option must be investigated.

Above all else, nothing should be declared as waste until it certainly is and don't be afraid in the meantime to admit you don't have perfect foresight about the future.

Nuclear waste: what's happening now?

What progress has been achieved in resolving the waste issue? It remains clear that this remains the major Achilles' heel of the nuclear industry in gaining popular public acceptance, but this has essentially been self-inflicted. As soon as the industry accepted the proposition that used nuclear fuel should be categorised as waste, a significant problem for public acceptance was created. If used fuel is indeed waste, a solution has to be found for its disposal that can be applied within the current generation – it cannot be passed on as a future liability for subsequent generations to deal with, as this contravenes the concept of sustainable development. Yet the reprocessing of the used fuel has always been a possibility and has been implemented in practice in a number of countries. Regarding used nuclear fuel as a potential asset with valuable materials to be recovered for future recycling in reactors is a different approach which gives the industry a lot more time in developing solutions.

Having conceded the crucial point that used fuel is waste, the industry has then tried to say that there isn't much of a problem anyway, as the volume of high level waste is so small. It has then confused the issue by building in concepts of 'retrievability' into repositories, just in case the material either causes unforeseen problems to the environment or suddenly acquires economic value. But what's the point spending billions of dollars putting this stuff down a hole in the ground when you're maybe going to bring it back to the surface again? In fact, the general public finds the once through cycle with deep repositories difficult to accept. If the stuff has to be put so far below ground, it must be really nasty and therefore is not something we want to pass onto the next generations. It is not really final disposal – the problem essentially remains.

There are four ideas for getting out of the mess that the industry has created for itself. Firstly, don't be afraid to admit that you don't know whether used fuel will be an asset or liability, as you can't be certain what future nuclear fuel markets will look like or how technology will shift. So try to promote the concept of long-term surface storage to the public on the basis that you are passing a potential asset onto the next generation, not a certain liability. Secondly, continue to investigate and demonstrate the technical merit of deep repositories as, whatever occurs, some of these are going to be needed in the future. Thirdly, look positively at the concept of international repositories. There are significant regulatory (and maybe public acceptance) problems with these, but the idea of each nuclear country, however small, having its own looks ludicrous from several angles. Finally, actively pursue research in improved technology in reprocessing, which should maybe take place a limited number of safeguarded sites around the world (as has also been suggested for enrichment facilities).

It is clear that there has been significant movement within the industry over the past 18 months on at least the first and last of these. There is now increased interest in reprocessing used fuel, following a period of interim storage, as opposed to the previous strong rush to establish repositories.

In the United States, reprocessing was banned by President Carter in the 1970s owing to concerns about the potential misuse of plutonium. Current reprocessing technologies in use throughout the world also create a significant amount of low- and intermediate-level nuclear waste, which create public acceptance problems. Yet for several years, there has been interest in new forms of reprocessing which do not separate plutonium from uranium (in fact recovering both together), and which segregate other actinides from fission products, enabling the actinides to be burned. The US budget process for 2006 included $50 million to develop a plan for 'integrated spent fuel recycling facilities', and a programme to achieve this with fast reactors will be a major US budget request in the future. As with reprocessing elsewhere, a large part of the incentive is to reduce volumes of high-level wastes and simplify their disposal. This doesn't, however, mean that waste repositories such as Yucca Mountain will never be needed – they must still be planned for and developed, but the quantities of material destined for them will be much reduced.

The difficulties encountered with establishing Yucca Mountain as an operating repository have undoubtedly influenced the move towards advanced reprocessing technologies. The likelihood of having to establish several Yuccas in the United States alone, if there is a significant boom in nuclear power in the 21st Century, is also important. Of lesser importance, but still relevant, is the recent strong upward movement in world uranium prices. Although the nuclear industry is convinced that there are more than adequate uranium reserves and resources to fuel any conceivable growth path of nuclear energy this century, the higher prices which are likely to be necessary to develop all the new mines will make recycling uranium and plutonium from used fuel relatively more attractive in an economic sense. When uranium prices were depressed by the ready availability of secondary supplies, there was a widespread perception that uranium would be very cheap forever, making recycling hard to justify. There is now recognition that new uranium mines require substantial capital inputs which must be recovered by adequate prices, also giving a fair return to the mining company.

Moves towards advanced fuel cycles in conjunction with new-generation reactors are likely to lead to significant re-evaluation of the fate of used fuel from present light water reactors and those about to be built. Some countries might find that moving to new fuel cycles, initially storing their light water reactor fuel and later re-using it, could be attractive. There seems to be a shift in attitudes about the value of used fuel that could eventually have repercussions for many national waste management programmes. Some facilities currently envisaged as final disposal repositories may only be used for interim storage of spent fuel that will eventually be reprocessed and recycled, hence the trend towards retrievability. Provision of a long-term storage service, possibly linked to fuel reprocessing and regeneration services, could be of great interest to some, while others may continue to prefer simply to dispose of used fuel.

There is also a connection with the International Atomic Energy Agency (IAEA) proposals on the creation of multinational fuel cycle facilities. Although the primary motivation may be to reduce the weapons proliferation risk and the immediate focus is very much on enrichment facilities, the idea also includes reprocessing facilities and repositories in the 'back end' of the nuclear fuel cycle too. One concept is to have a limited number of advanced reprocessing facilities and waste repositories in a small number of countries, but under multinational control, with guarantees of security of supply to nations willing to fulfil full safeguards obligations. This may well include the possibility of fuel leasing, whereby fuel is supplied to reactors and then taken back for reprocessing and/or disposal, without the utility that owns the reactor ever taking ownership of the fuel.

With these developments, it is possible to look forward to an even lower fuel price element in the economics of Generation IV reactors. Low and relatively stable fuel prices are already a significant advantage of the current generation of evolutionary reactors against alternative fossil fuel generating modes, but the future looks even better. There are already significant quantities of separated civil plutonium, reprocessed uranium and depleted uranium in inventory and these may well be utilised when the new reactor designs become reality. It is almost certain that fresh uranium will still have to be mined, but the quantities will be much lower than required by the current generation of reactors.

So far as public perception of nuclear power is concerned, it will undoubtedly be difficult to shift opinion towards the idea that the industry does not create large quantities of dangerous wastes. Moving towards longer-term storage of used fuel with the expectation that it will eventually be reprocessed should help in this but it is important to demonstrate that the industry is not just "passing the buck" to the next generation. Used fuel must somehow be presented as an asset, as a key foundation for fuelling the next generation of reactors, without the need to mine and utilise greater quantities of a finite resource, such as uranium, than is really necessary. Over two million tonnes of uranium have been mined since 1945, both for military and civil nuclear programmes, and most of it is readily identifiable today in the form of depleted uranium from enrichment plant operations. It makes sense to use as much as possible in the future of what were formerly regarded as wastes from previous nuclear operations as true fuel assets in new reactor types.

Reprocessing used fuel: devil or saint?

The advantages and disadvantages of used nuclear fuel reprocessing have been debated since the dawn of the nuclear era. There is a range of issues involved, notably the sound management of wastes, the conservation of resources, economics, hazards of radioactive materials and potential proliferation of nuclear weapons. Sifting through these is not easy, with strong counter-claims made by opposing parties, but it is undoubtedly true that in recent years, the reprocessing advocates appear to be winning once again, perhaps most clearly demonstrated by the apparent change in position of the United States under the Global Nuclear Energy Partnership (GNEP) programme.

Reprocessing advocates claim that the closed fuel cycle approach is the best way to support a sustainable life for nuclear energy with ecological responsibility, as it reduces recourse to natural uranium resources and optimises waste management. As a start, it's important conceptually to distinguish reprocessing from recycling. Reprocessing is stage one – the separation of uranium and plutonium out of used fuel and conditioning of the remaining material as waste. Fuel assemblies removed from a reactor are very radioactive and produce heat, so are cooled (mostly at the reactor site or otherwise at a central storage facility or at the reprocessing plant) for a number of years as the level of radioactivity decreases considerably. For most types of fuel, reprocessing occurs anything from 5 to 25 years after reactor discharge. Recycling is then stage two – the use of the uranium and plutonium from the reprocessing plant, which can be either as mixed oxide (MOX) fuel or reprocessed uranium (RepU) fuel in current reactors or as fuel for future Generation IV reactors. Reprocessing effectively sets up the possibility of recycling. This doesn't necessarily have to follow, but in practice, the two stages are bound together as reprocessing will likely only be undertaken with a view to eventual recycling.

Although world uranium resources are extensive, many people do not see them alone as a lasting solution to world energy needs. If nuclear power expands rapidly, the quantity of uranium mined and processed each year may have to rise from the current 40,000 tonnes per annum, to 100,000 tonnes and beyond. The management of nuclear waste remains one of the main concerns of the public and could also constrain the future expansion of nuclear. For nuclear power to be seen as sustainable, it is important that natural uranium resources are conserved as much as possible and that wastes are managed in a safe and durable way. Reprocessing arguably contributes to sound stewardship of uranium resources by allowing the recycling of reusable materials, which introduced as MOX fuel and RepU can save around one quarter of uranium needs. It also minimises high-level waste toxicity and volume; after treatment, the level of toxicity is only one tenth of what it previously was and its volume down to one fifth. The wastes can be conditioned into a passive form that will be safely stored pending final disposal. The scientific and technical community generally feels confident that there already exist technical solutions to such used fuel and nuclear waste conditioning and disposal, with a wide consensus on the safety and benefits of geological disposal.

A great deal of reprocessing has been going on since the 1940s, originally for military purposes, to recover plutonium for weapons (from low burn-up used fuel, which has been in a reactor for only a very few months). So far, some 80,000 tonnes (of 280,000 tonnes discharged) of used fuel from commercial power reactors has been reprocessed.

In the UK, metal fuel elements from the first generation of gas-cooled commercial reactors (Magnox) have been reprocessed at Sellafield for about 50 years. The 1500 tonnes per year (t/y) plant has been successfully developed to keep abreast of evolving safety, hygiene and other regulatory standards. Some 15,000 tonnes of RepU from Magnox reactors has been used for enriched advanced gas-cooled reactor (AGR) fuel. From 1969 to 1973 oxide fuels were also reprocessed, using part of the Magnox reprocessing plant modified for the purpose. The 900 t/y Thermal Oxide Reprocessing Plant (THORP) was commissioned in 1994 and the corresponding mixed oxide (MOX) fuel plant in 2001.

In France one 400 t/y reprocessing plant was operated for metal fuels from gas-cooled reactors at Marcoule. More significantly, however, at La Hague, reprocessing of oxide fuels has been carried out since 1976, and two 800 t/y plants are now operating, while there is also the Melox MOX fuel fabrication facility at Marcoule. Currently, the reprocessing of 1150 tonnes of EDF used fuel per year produces 8.5 tonnes of plutonium (immediately recycled as MOX fuel) and 815 tonnes of RepU. EDF has demonstrated the use of RepU in its 900 MWe power plants, but it is has been uneconomic during the years of low uranium prices, due to conversion costing significantly more than for fresh uranium, and enrichment needing to be separate because of U-232 and U-236 impurities in the reprocessed fuel. Provision has been made, however, to store reprocessed RepU for up to 250 years as a strategic reserve.

Elsewhere in the world, India has a 100 t/y oxide fuel plant operating at Tarapur, while Japan is currently commissioning a major (800 t/y) reprocessing plant at Rokkasho, having had most of its used fuel previously reprocessed in Europe at La Hague and Sellafield. Russia has a 400 t/y oxide fuel reprocessing plant (Mayak) at Ozersk (Chelyabinsk). In the USA, no civil reprocessing plants are now operating, though three were originally built. The final one, at Barnwell in South Carolina, was closed in 1977, owing to a change in government policy, ruling out all civilian reprocessing as a facet of US non-proliferation policy.

Much of the opposition to reprocessing centres on a huge amount of hype about plutonium created by anti-nuclear forces. Talk of dependence on a 'plutonium economy', if nuclear power expands much further, is still common in their publications. Yet despite being toxic both chemically and because of its ionising radiation, plutonium is far from being "the most toxic substance on Earth" or so hazardous that "a speck can kill". On both counts there are substances in daily use that, per unit of mass, have equal or greater chemical toxicity (such as arsenic and cyanide) and also radiotoxicity (the plutonium decay product americium-241 used in smoke detectors).

Plutonium is one among many toxic materials that have to be handled with great care to minimise the associated but well understood risks. Thousands of people have worked with plutonium, with their health protected by the use of remote handling, protective clothing and extensive health monitoring procedures. The main threat to humans comes from inhalation, where it can be trapped and readily transferred, first to the blood or lymph system and later to other parts of the body, notably the liver and bones. It is here that the deposited plutonium's alpha radiation may eventually cause cancer. Nevertheless, the hazard is similar to that from any other alpha-emitting radionuclides that might be inhaled. It is less hazardous than those that are short-lived and hence more radioactive, such as the decay products of radon gas, which (albeit in low concentrations) are naturally common and widespread in the environment.

The other aspect of plutonium which raises the ire of nuclear opponents is its alleged proliferation risk. Opponents of the use of MOX fuels commonly state that such fuels represent a proliferation risk because the plutonium in the fuel is said to be 'weapons-useable'. But MOX is a mixture of uranium and plutonium oxides, with the plutonium being very much in the minority. For light water reactor fuel, the plutonium content is typically around 5%. MOX cannot possibly be used in nuclear weapons or nuclear explosives.

To separate the plutonium content from MOX fuel elements would be a major undertaking, similar to reprocessing.

In addition, there would be serious technical difficulties in attempting to make nuclear weapons from plutonium of the quality currently used for MOX and none of the countries possessing nuclear weapons has ever made weapons using plutonium of this quality. Rigorous International Atomic Energy Agency (IAEA) safeguards also apply to this material in non-nuclear-weapon states party to the Treaty on the Non-Proliferation of Nuclear Weapons (NPT). It is misleading to conclude that, because this material is subject to safeguards, it is therefore 'weapons-useable'. The plutonium isotope most suitable for weapons use is Pu-239, which comprises at least 92% (and usually more) of a nuclear weapon. This plutonium is produced in dedicated plutonium production reactors, specially designed and operated to produce plutonium of this quality by removal and reprocessing of fuel after short irradiation times. The plutonium produced in the normal operation of light water reactors, from which MOX fuel is being made, has a substantial proportion of higher plutonium isotopes, so that it typically comprises less than 60% of Pu-239. The remainder contains a large proportion of isotopes which create serious technical difficulties for weapons use, namely Pu-238, Pu-240 and Pu-242.

Another criticism of reprocessing centres around the transportation of materials, notably the shipments of used fuel from Japan and other countries to the reprocessing facilities in France and the UK and the subsequent return of MOX fuel and high-level wastes. This is very much a special case of the difficulties of transporting any fissile materials today but, in reality, involves no greater risk to human health and to weapons proliferation. The sea shipments have been made by purpose-built vessels and involve special casks that have been subject to rigorous impact testing. The only risk to human life has come from the actions of protesters who have regarded the shipments as visible symbols of technologies with which they remain uncomfortable.

Finally, the reprocessing facilities themselves have come under attack through alleged risk to the local population through contamination of the air and of nearby beaches and the possible risk to people much further away by contamination of the sea. Studies have so far failed to show any conclusive risk to either the workforces or the local population – the leukaemia clusters found near Sellafield in the UK have been explained by alternative causal factors. Both French and UK plants, however, have stringent targets for reducing the concentrations of radionuclides in the nearby seawater, although it is not accepted that the current or previous levels are in any way harmful to human health.

In summary, the reprocessing of used fuel should not give rise to any particular public concern and offers a number of potential benefits in terms of optimising both the use of natural resources and waste management. Whether it will become the generally accepted way of dealing with used nuclear fuel remains to be seen, as several major countries are still very much attached to a once through fuel cycle, with used fuel sent directly to a repository.

Recycling uranium and plutonium: where's it heading?

The availability of recyclable fissile and fertile materials able to provide fresh fuel for existing and future nuclear power plants is a key characteristic of nuclear energy. Programmes for the recycling of plutonium were developed in the 1970s when it appeared that uranium would be in scarce supply and would become increasingly expensive. It was originally proposed that plutonium would be recycled through fast breeder reactors, that is, reactors with a uranium 'blanket' but which would produce slightly more plutonium than they consume. Thus it was envisaged that the world's 'low cost' uranium resources, then estimated to be sufficient for only 50 years' consumption, could be extended for hundreds of years.

As things transpired, the pressure on uranium resources was very much less than expected and prices remained low in the period up to 2003. This was caused by the discovery of several new extensive and low-cost uranium deposits, the entry onto the world market of large quantities of uranium from the dismantling of nuclear weapons and the slower growth of nuclear power than was expected back in the 1970s. There became little incentive to develop fast breeder reactors, particularly as these present major engineering challenges, which could prove expensive to resolve. Nevertheless, since the late 1970s, around 30% of spent fuel arisings from commercial nuclear reactors outside the former Soviet Union and its satellite states have been covered by reprocessing contracts with plants in France and the UK. Without fast breeder reactors, there has been an accumulation of separated plutonium stockpiles.

Mixed oxide (MOX) fuel was introduced mainly to reduce the stockpiles of plutonium, which were building up as spent fuel reprocessing contracts were fulfilled. MOX was therefore an expedient solution to a perceived problem, which had been created by changed circumstances. The MOX programmes have demonstrated that plutonium has some advantages as a nuclear fuel and so the stockpiles have economic value. The MOX era, however, may pass relatively quickly, even if plutonium stockpiles worldwide are not substantially reduced. Revived interest in nuclear power in the 21st Century, as a clean air solution which contributes to world sustainable development, is encouraging the development of new materials and technologies. In addition, the substantial rise in uranium prices since 2003 and the difficulties with commissioning waste repositories have prompted the beginning of a revaluation of recycling.

Currently 12 of the countries with nuclear energy programmes are committed to a closed nuclear fuel cycle but there are signs that the number will soon increase. In particular, the United States is reassessing its previous policy, which was set strongly against reprocessing with subsequent recycling of recovered materials. The decision to introduce MOX fuel from ex-weapons plutonium in civil reactors was an important element in this and the first assemblies are now in use in reactors operated by Duke Power. In November 2005 the American Nuclear Society released a position statement saying that it "believes that the development and deployment of advanced nuclear reactors based on fast neutron fission technology is important to the sustainability, reliability and security of the world's long-term energy supply." This will enable "extending by a hundred-fold the amount of energy extracted from the same amount of mined uranium." The statement envisages onsite reprocessing of used fuel from fast reactors and says that "virtually all long-lived heavy elements are eliminated during fast reactor operation, leaving a small amount of fission product waste which requires assured isolation from the environment for less than 500 years."

The Global Nuclear Energy Partnership (GNEP) programme, announced by the US Department of Energy in early 2006, fits in closely with this. A major issue addressed is the efficiency of the current nuclear fuel cycle. The 'once through' cycle only uses part of the potential energy in the fuel, while effectively wasting substantial amounts of useable energy that could be tapped through recycling. While European countries and Japan have recycled some of the residual uranium and plutonium recovered from the spent fuel in light water reactors through MOX utilisation, no one has yet employed a comprehensive technology that includes full actinide recycle. In the United States this question is pressing since significant amounts of used nuclear fuel are stored in different locations around the country awaiting shipment to the planned geological repository at Yucca Mountain in Nevada. This project is much-delayed, and in any case will fill very rapidly if it is used simply for used fuel rather than the separated wastes after reprocessing it.

An early priority in GNEP is therefore the development of new reprocessing technologies to enable recycling of most of the used fuel. One of the concerns when reprocessing used nuclear fuel is ensuring that elements separated are not used to create weapons. The Purex process, used in all existing reprocessing plants, has been employed for over half a century and has resulted in the accumulation of 240 tonnes of separated reactor-grade plutonium around the world (though some has been used in MOX). While this is not viable for weapons use, it is no longer seen as appropriate and future reprocessing will result in the plutonium being combined with some uranium and possibly with minor actinides. GNEP creates a framework where states that currently employ reprocessing technologies can collaborate to design and deploy advanced separation and fuel fabrication techniques that do not result in the accumulation of separated pure plutonium.

Several developments of Purex which fit the GNEP concept are being trialled in different countries, notably Urex+ in the United States and Coex in France. The latter separates uranium and plutonium (and possibly neptunium) together as well as a pure uranium stream, leaving minor actinides with the fission products. The central feature of these variants is to keep the plutonium either with some uranium or with other transuranics which can be destroyed by burning in a fast neutron reactor – the plutonium being the main fuel constituent. Trials of some fuels arising from Urex+ reprocessing in the United States are being undertaken in the French Phenix fast reactor.

The second main technological development envisaged under GNEP is the advanced recycling reactor – basically a fast reactor capable of burning minor actinides. Thus used fuel from light water reactors would be reprocessed at a recycling centre and the transuranic product transferred to a fast reactor onsite, which both produces electricity at a capacity of perhaps 1000MWe and incinerates the actinides. A key objective of this programme is to obtain design certification of a standard fast reactor from the US Nuclear Regulatory Commission. Related to this, nearly all the new reactor models being developed under the Generation IV and the International Project on Innovative Nuclear Reactors and Fuel Cycles (INPRO) projects have closed fuel cycles recycling all the actinides. Although part of the motivation remains making savings in the use of the (now more expensive again) natural uranium resource, the key today is saving on used fuel arisings and developing ways to deal with the existing volumes of fuel, created by commercial nuclear power to date.

Looking at the more immediate term, EDF will continue to send for reprocessing 850 tonnes of its 1,200 tonnes of used fuel discharged each year. The remainder is preserved for later reprocessing to provide the plutonium required for the startup of Generation IV reactors, the prototype of which is envisaged by 2020. In Japan, the Rokkashomura reprocessing plant should be commissioned in mid 2008. European reprocessing of Japanese used fuel ended in 2005 and it is envisaged that 16-18 reactors will eventually be loaded with MOX fuel. Aomori prefecture in 2005 also approved construction of a MOX fuel fabrication facility but this is not expected to open until 2012 at the earliest – until then, fabrication of MOX for Japan will take place in Europe.

In the United Kingdom, the plant reprocessing Magnox fuel will close in 2012, following the permanent shutdown of all the reactors it serves and there continue to be uncertainties surrounding the future of the THORP reprocessing plant and the associated MOX fuel fabrication facility. Nevertheless, the UK has a considerable inventory of separated reactor-grade plutonium (over 100 tonnes) as well as a substantial (about 60,000 tonnes) quantity of depleted uranium. This could form the foundation of an advanced reactor programme but a Royal Society report in September 2007 recommended that the plutonium be used in MOX fuel. This will depend on persuading reactor operators in the UK (including those running any new reactors) to adopt this as a fuelling strategy – it is by no means certain that they will.

Russia may eventually achieve its stated aim of closing its fuel cycle, although it has so far achieved very little in this direction. Plans for expanding the Mayak reprocessing facility or building a second plant, as well as a fuel fabrication facility, have so far come to nought, but the revival in the Russian nuclear industry and its interest in playing a similar role to what the United States envisages for itself in GNEP, suggest that there will soon be some new developments.

Finally, the strong upward movement in uranium prices suggests that utilities owning inventories of reprocessed uranium (RepU) will look once again at utilising these. The greater expense at the conversion and enrichment stages may now be outweighed by the substantially increased prices for fresh fuel. EDF is at centre stage here, owning significant quantities of RepU as a strategic asset. A few years ago, these could fairly be viewed on the other side of the balance sheet, as a long term liability, but such an assessment is now outdated. Certainly many European utilities (and maybe also some in the United States) are looking at RepU in a new light and possibly seeking to add to those who have already gone down this road (albeit in relatively small quantities).

To summarise, it seems clear that recycling remains a very live issue in the nuclear sector, indeed with an apparent push from several quarters to pursue it more vigorously in the future. Used fuel management is a huge and still growing business and options are being sought that meet a variety of requirements, certainly not merely economic but also considering environmental, resource sustainability and non-proliferation objectives.

Decommissioning activity: the sunset for nuclear?

Decommissioning of nuclear facilities is now big business. Many uranium mines, commercial nuclear power reactors, research reactors and other fuel cycle facilities have already been retired from operation. Some of these have now been fully dismantled and sites returned to alternative uses, while others await this in the future. The sums of money involved are considerable and the business is very attractive to both major international contracting companies and smaller local specialists. One sign of this is the number of business conferences devoted to decommissioning – commercial conference organisers always react quickly to obvious new opportunities.

Those opposed to nuclear power have seized on the current extent of decommissioning activity as an additional weapon to use against the industry and in particular a revival in the construction of new reactors. They point to the sums of money involved in cleaning up old nuclear sites and assert that a similar experience will follow in the future if new reactors are constructed. The implication is that not all costs are being incorporated when economic evaluations are made of new reactors.

There is also the assertion that the extent of decommissioning activity marks a sunset for the nuclear industry. The switch of activities away from new reactor build towards shutting down and cleaning up existing sites is seen as demonstrating that a mature industry has now moved towards its final death.

The sums of money involved are certainly considerable. The £70 billion quoted in the UK as the cost of cleaning up the historical legacy of nuclear liabilities represents over £1000 for every UK citizen. However, this is to be spread over a considerable period of time in the future and will be spent under competitive tendering by the new Nuclear Decommissioning Authority (NDA). The UK has at least bitten the bullet by stating a figure but the total sums of money required in the United States and the former Soviet Union are likely to be considerably in excess of this. It can certainly be viewed as a burden left to future generations by those in the recent past but the annual sums involved are not huge in comparison with public expenditure on defence, education, health or even road-building. There are no obvious implications for public health and safety in cleaning up the sites over a considerable period of time, so it is best to spread the burden and also to profit from the increased experience of learning by doing.

The key point, however, is that these nuclear sites mostly date from the early days of nuclear research and development (R&D) in the 1950s and 1960s and cannot fairly be laid at the door of the commercial nuclear power industry today. At this time, the thrust of R&D was very much towards military applications with commercial nuclear power merely a sideshow. Far less attention was paid to the environmental consequences of the activities than would be acceptable today. We are now subjecting historical decisions to new, stricter modern rules and regulations. This is akin to the industrial sites from the Industrial Revolution that blighted the UK landscape for many years, most in close proximity to major centres of population. Also the industrial operations in what are now National Parks. Nobody cared too much at the time for the mess created but society has subsequently coped well with cleaning up and reclaiming the sites for fresh uses. The nuclear clean-up legacy, although considerable, can perhaps be put down as an additional cost of the Cold War, which must be covered by future generations who are now benefitting from the maintenance of political freedom.

Any new nuclear build will have to incorporate adequate decommissioning funding in its financial analysis. In fact, the sums of money that must be put away on an annual basis are not material to the economic viability of new nuclear reactors in comparison with competing technologies such as coal- and gas-fired generation. Decommissioning funds are rather like personal pension funds – if relatively small amounts of money are put away in the early days, the nature of compound interest means that after 50 or 60 years, substantial funding is available to meet the liability. Most modern reactor designs are also simpler and smaller in scale of construction than the older reactor types, so should be easier to decommission after their working lives. The experience gained of nuclear decommissioning should also be considerable by then, meaning that costs should be falling.

Turning to the additional argument presented by those opposed to the nuclear industry, is it now past maturity or still lying at an early stage of development? Certainly, the lack of new reactor construction in most countries is hardly a sign of robust health. Only in India and China can the prospects be described as exciting, with only a few other bright spots in evidence around the globe, such as the fifth Finnish reactor. The lack of new orders poses significant problems for maintaining the capacity to keep the industry going in the future. The closure of university courses in nuclear engineering and reactor physics is a sign that young people are not being attracted to a career in nuclear. The attention that the industry has given to this phenomenon shows the seriousness with which it is regarded. The establishment of the World Nuclear University marks recognition that the industry needs to develop its future leaders, at a time when opportunities are perhaps more obvious elsewhere to young people.

Taking the UK as an example, the government's historically weak 'support' for the industry ("keeping the option open") and the switch of activities within BNFL towards decommissioning and clean-up through acting as a contractor to the NDA would seem to indicate that the sun is going down on the industry and is in imminent threat of setting. Although it has long been hoped that the advanced gas-cooled reactors (AGRs) could potentially run longer than their current licences, technical issues now suggest that this is unlikely and by 2023, only the Sizewell B PWR will still be in operation. The reprocessing of used fuel at Sellafield also has an uncertain future beyond 2010. The only expanding part of the nuclear industry in the UK, with the exception of waste management and decommissioning, is uranium enrichment by Urenco at Capenhurst, where capacity is expanded on a modular basis as new contracts are signed.

It is possible, however, to paint a much more positive picture. If you believe that a substantial nuclear revival will be required for sound economic and environmental reasons, the period between the mid 1980s and today can be seen as a mere hiccup in the longer-term development of the industry. The accidents at Three Mile Island and Chernobyl caused the industry to pause for breath as its safety and economics came under closer scrutiny. The extent of clean-up activities from the experimental early days of nuclear is a one-off expense which will never be repeated – beyond that, the decommissioning today of early reactors, mines *etc* is an entirely normal activity where the experience gained will earn rewards for the future. Looking many years into the future, when Generation IV reactors will be operable, nuclear power may play a considerable role in hydrogen production, and seawater desalination, in addition to power generation. The late 20th and early 21st Centuries may then be seen as nuclear's own pause for breath and period for a reality check.

The existing stock of nuclear reactors, amounting to 439 around the world, is also likely to last considerably longer than many outside critics and commentators suggest. Apart from early reactors in the UK and a few other countries, the only reactor closures are likely to be those which are politically-inspired in countries such as Sweden and Germany and some of old Russian designs in Eastern Europe. Elsewhere, operating life extension of reactors is an accepted, highly economic method of meeting rising electricity demands. When reactors can be kept going longer at acceptable costs, delaying the employment of decommissioning funds is an additional reason for carrying on operating for additional years.

In fact, the nuclear industry shows many other signs of immaturity when compared with other industries. It is still treated like an 'infant industry' in many countries, with large amounts of government control and state intervention. Reactor design and supply are still undertaken by a large number of companies, whereas a more mature industry would show greater signs of rationalisation towards a small number of suppliers. There is greater rationalisation evident elsewhere in the nuclear fuel cycle, where uranium supply, conversion and enrichment are all now largely in the hands of large, highly economic operators. Yet the nuclear fuel market is far from mature, relying on a rather haphazard mix of long-term contracts and a creaky spot market.

It must be hoped that decommissioning of facilities will gradually be accepted as an everyday part of the nuclear industry and not something that attracts particular note or comment. The historical clean-up liabilities will eventually be dealt with and more modern facilities financed in a way that clearly covers all possible costs. It is important that the industry convinces its opponents that all costs are incorporated, particularly as it points the finger at the non-incorporation of external costs of fossil fuel-fired generation modes.

6. THE BIG PICTURE

Introduction World energy matters have now returned as a subject of popular debate, after many years on the sidelines after the world oil crises of the 1970s. This has been prompted by renewed concerns over the security of long-term oil and gas supplies (indicated by significant price escalation) but also by the concerns about the environmental consequences of continued mass exploitation of fossil fuel resources. Much improved world economic performance in recent years has heightened both of these concerns and notably the dramatic growth rates now being achieved in the largest developing countries, China and India.

Within this, there are various scenarios postulated for the future of nuclear power. Some observers doubt if the stranglehold of fossil fuel exploitation can easily be reversed and see this continuing through most of this century. Increasingly, however, alternative strategies are being advocated, which involve renewable energy sources and also a revival of nuclear power. The International Energy Agency's (IEA's) work highlights the difficulties of continuing with fossil fuels in an era of rapidly-expanding energy use up to 2030 but also demonstrates that it is far from easy to break with the recent past and develop superior solutions. The way is clearly open, however, for nuclear power to demonstrate that it can play the same important role in world energy envisaged back in the 1970s – the intervening years may eventually be seen as a mere hiccup.

The nuclear industry itself is not really the strong monolith that is sometimes depicted by its opponents. Its structure is somewhat fragmented, at least by comparison with other major industries, notably oil and gas, which is at least partly explained by its complexity. Nevertheless, as its natural development got cut off at some point in the 1980s and for many of the years since it has been essentially 'marking time', only one major integrated company has ensued, in the shape of Areva in France. This may now change as a strong nuclear revival takes place, with further consolidation and integration of industry players. Industry representation by professional and lobbying organisations will also likely consolidate, to become more effective. Nevertheless, the nuclear industry has some special features that will always make it somewhat different from others. But, as far as possible, it should try to present itself as not particularly different and try to act as such without too many government restrictions on production and trade.

Lifecycle analysis (LCA) of various electricity generation options shows nuclear in a very favourable light, both in terms of its net energy contribution and its low external costs. The excellent operating performance of existing nuclear power plants demonstrates that they produce a huge amount of electricity from a small initial fuel input and without causing any substantial environmental disturbance of other impacts on human life. Technological improvements, such as the move to centrifuges and (possibly) lasers in the fuel enrichment stage are set to boost this further, as significant energy savings will be made, while in situ leach (ISL) mining allows low grade deposits to be developed with little surface disturbance.

Nuclear proliferation, however, remains as a very live issue and could conceivably threaten prospects of a nuclear revival. In reality, however, the international non-proliferation regime has been extremely successful (in that fewer than predicted countries have developed nuclear weapons) although it now looks a little frayed around the edges. Certainly ways need to be sought to bring India and Pakistan fully within the framework and avoid other countries going down the same road. This is being addressed in various ways, notably by initiatives such as the Global Nuclear Energy Partnership (GNEP) as proposed by the United States and the other, largely complementary, suggestions by other countries. Although these ideas have to overcome some objections based on concerns over infringing national sovereignty and commercial competition, they are very much in the right direction. An expansive nuclear sector vitally needs a lot of fresh thinking on the necessary international regimes, in order to cope with changed circumstances in the world today.

World energy in the 21st Century: how much nuclear?

We spend much of our time worrying about the present and the immediate future but it is interesting to sometimes cast one's thoughts further ahead. Energy is very much in the news these days, with the interest in global warming, oil and gas price escalation, the threat to security of supply and the attention cast on the renewable sources. What are anticipated to be the predominant energy trends throughout the 21st Century and how does nuclear power fit into the picture?

Most energy forecasts go to only 2030, notably those of the respected International Energy Agency. A 25-year forecasting period is already ambitious, given the inability of many predictions for even a few years ahead to be accurate. For example, *The Economist* magazine hates to be reminded of its cover story from March 1999 entitled *Drowning in Oil*, which predicted oil prices were "headed for

$5 per barrel." So it's always best to be humble. Yet forecasters occasionally venture out further ahead, right to the end of the century. They recognise that different techniques are required here, as so much can change, so usually talk in terms of alternative scenarios. For example, the Intergovernmental Panel on Climate Change (IPCC) utilises a large number of alternative energy scenarios to judge their impact on global warming.

Looking as far as 2100, it is useful to characterise two extreme positions on where energy supply and demand will go. Advocates of one side believe that things will change only slowly and there will be no dramatic shifts whereas the alternative stresses that change will be more dramatic, forced by rising shortages of energy resources or environmental concerns. It is useful to examine these two extremes – reality could no doubt lie somewhere in between but the limits are interesting.

Those believing that change will be evolutionary, not revolutionary, believe that fossil fuels will remain an important (and indeed essential) part of the energy mix throughout the 21st Century. As far as demand is concerned, they emphasise the slowing of energy demand growth once countries reach a certain level of economic development, as less energy-intensive sectors predominate. They also believe that world population will grow less quickly than some forecasters, also that higher prices for energy will help curb demand.

On supply, they make the (good) point that consistent talk of the world running out of oil has been proved wide of the mark and that exploration and production respond dynamically to economic criteria. They believe that coal will remain an important part of the energy mix in the first half of the century, as both China and India must make use of their abundant reserves to underpin their economic growth strategies. Oil supply will not peak until 2030 or 2040 because reserves are usually revised upwards over time and non-conventional reserves (from tar sands and the like), will become important. The latter depends on the oil price remaining firm, as production costs are substantially in excess of those of current wells.

The biggest feature of the century, however, will be the rise of natural gas, both through pipelines and in the form of liquefied natural gas (LNG). This is essentially a continuation of the recent trend where gas has risen sharply as a component of world energy supply. Indeed, this group characterises the 19th Century as the coal era, the 20th Century as the oil era and the 21st Century as the period of natural gas domination. Renewables only become important in the second half of the century, as technology advances and their economic competitiveness improves. Nuclear is largely seen by this group as an aberration; an unwarranted panic response to the oil price hikes in the 1970s and 1980s, which were unsustainable as the market adjusted. So nuclear plants are expected gradually to close down throughout the century and there will be few new ones – so by 2100 nuclear power will be no more.

This group is doubtful that global warming will be a trigger for substantial change in energy supply. This is not necessarily because they doubt the postulated link between carbon emissions and warming – rather that they believe that energy demand will rise much more slowly than many forecasters (particularly the IPCC scenarios) while the gradual substitution of coal and oil by natural gas and renewables is sufficient to minimise the worst effects.

The alternative view sees more dramatic and revolutionary change, based on forecasts of likely depletion of fossil fuel resources and their adverse environmental impact. Many accept the thesis that global warming is, indeed, the biggest problem affecting mankind in the 21st Century and that all possible measures must be taken to mitigate this. The use of all fossil fuels, including natural gas, will be curtailed by carbon taxes, emissions trading schemes and other measures, and renewable sources will rise rapidly to meet the gap. The use of coal, oil and gas will be gradually replaced by the rise of hydrogen as a low-emissions fuel. This will become necessary as demand, led by the developing countries, will rise quickly throughout the century, increasing the imperative to do everything possible to cut carbon emissions.

The whole energy supply infrastructure, built up in the second half of the previous century, will need replacing. The existing oil product distribution system to meet the demands of the transportation sector will need to be replaced by one adapted to delivering hydrogen where and when it is needed. Electricity supply will become decentralised with generation taking place more locally, with many small units contributing to the grid, including possibly households with small heat and power units. Large generating units, with their substantial transmission losses, will gradually become a thing of the past.

Nuclear may or may not find a substantial place in this world. Some would argue that it is unwanted and unnecessary because a combination of renewables and hydrogen will be sufficient to meet demand and the public acceptance and waste problems will never be overcome. Others claim that nuclear is essential as renewables cannot be expected to take up the slack from fossil fuels until well into the century, while nuclear reactors can also play a major part in hydrogen production.

The two extreme positions on 21st Century energy are clearly poles apart, but nuclear has an uncertain future in both. It is essentially dismissed in the first and may or may not play a role in the second. It is possible, however, to see a better place for nuclear somewhere between the two extreme views and particularly if we believe that the technology can move steadily forwards.

It is clear that the current generation of evolutionary light water reactors may be adequate to engineer a nuclear revival in the period to 2030, if capital costs can be cut and natural gas prices are high. Nevertheless, a new generation of nuclear reactors will be required to meet the demands of the remaining years of the century: more economical to build and operate, producing less waste and more proliferation-resistant. This is essentially what the Generation IV and International Project on Innovative Nuclear Reactors and Fuel Cycles (INPRO) projects are currently trying to do, by developing a range of designs that meet these key criteria.

If these can achieve much of what they promise, a substantial role for nuclear may even be found if the first of the extreme energy worlds turns out to be closer to the mark. Natural gas may become the fuel of the 21st Century, but it will have to compete with nuclear for power generation. Unless gas prices can remain low (which seems unlikely as more supply must come in the form of LNG, requiring substantial port infrastructure and transport costs) it can be argued that new nuclear plants will have a excellent chance of competing. Clean coal technology plants may be another option, which should not be ruled out, but the costs are uncertain and may not be competitive with nuclear.

If fossil fuels gradually lose their grip on energy supply, as the second case advocates, the chances of nuclear could be even better. The next generation of reactors have a good chance of overcoming many of the public perception problems that turn people off nuclear today. Doubts about how quickly renewables can expand will necessarily push attention towards nuclear as the only large-scale option that can replace fossil fuels. Nuclear may conceivably come to be regarded as a renewable source of energy as well as clean. Nuclear's possible role in hydrogen production may then additionally reinforce its position. It makes little sense to produce hydrogen with fossil fuels, as the net environmental benefit will be substantially reduced, particularly if it is coal that is used to fuel the plants.

In reality, the energy world is more likely to evolve as the complex mixture of the two extreme cases. Some of the demand scenarios developed by the IPCC, for example, seem rather on the high side, as they would bring many developing countries up to per capita energy consumption figures common in some of the more wasteful developed countries today. A more energy-conscious way of living should be possible by then. It is also unlikely that fossil fuels will have as clear a run as the first case believes, partly for geopolitical reasons (notably concerns about security of supply from unstable regions of the world) but also due to price pressure as resources become increasingly depleted.

So nuclear has an excellent chance in the period beyond 2030, provided that the Generation IV and INPRO reactors are developed in the envisaged way. Their development is going to require many years of substantial public funding, to be justified by the need to invest in the world's future energy supply infrastructure.

IEA World Energy Outlook: where is nuclear?

The *World Energy Outlook 2006* (WEO 2006) published by the International Energy Agency (IEA) represented a considerable advance for a number of reasons. This annual report has long been recognised as the most comprehensive one covering energy trends. Having been mandated by world leaders to report on the ways in which recent energy trends (widely regarded as unsustainable for environmental and security of supply reasons) can be changed at minimum cost and disruption, the report set out a detailed alternative policy scenario which goes a long way to achieving this. That this includes a significant prospective role for nuclear power has naturally been pleasing for the industry, but the force with which the IEA said this has taken many people by surprise.

The IEA has historically been regarded as at best agnostic about nuclear. Some have seen this as being as a result of its close identification with oil and gas interests – indeed, the IEA was set up in the aftermath of the early 1970s oil price escalation, as an Organisation for Economic Cooperation and Development (OECD) body. Its remit has always involved detailed analysis of the fossil fuels world, with little attention given to nuclear. As the IEA works essentially on an assumption of unchanged government policies, it has felt free to ignore nuclear during a period when it has been essentially out of favour in major governmental circuits. Its energy models also build in economic factors as important determinants of supply and demand, so with nuclear appearing to be a generally uneconomic option, it could easily be put to one side.

Now that nuclear is getting much more attention from governments around the world and its economic prospects have undoubtedly improved (quite apart from anything the industry has done to cut costs, the sharp increase in fossil fuel prices ensure this is true), the IEA cannot ignore it any longer. Indeed, the increased requirement for the IEA to come up with energy options that avert greenhouse gases has pushed it gradually towards nuclear over a longer period of time. Yet the IEA is not really making forecasts in its report – it is merely assessing what will be the outcome if present policies stay essentially unchanged (the reference scenario, RS) or are modified in particular ways (the alternative policy scenario, APS). As such, it is doing something rather different to the World Nuclear Association (WNA), whose nuclear generating capacity scenarios are also widely quoted. The WNA develops three distinctive scenarios, reference, lower and upper, each of which aims to be a realistic and believable view of the future, but containing different generic assumptions on much more than government policies – such as nuclear economics, responses to environmental concerns and movements in public acceptance.

One of the most pleasing aspects of WEO 2006 is that it dispels several of the myths that the anti-nuclear movement has tried to build up. For example, the claim that there will be insufficient uranium to fuel strong nuclear growth is comprehensively rebutted. Also the argument that nuclear cannot contribute in a substantial way to the avoidance of greenhouse gas emissions. The economic analysis is also revealing, showing that decommissioning costs, where provided for adequately, are a very minor consideration in new nuclear plant economics. The fact that nuclear costs, both fuel and operations and maintenance, are relatively low and stable over time, is also highlighted. But the best part is the demonstration that new nuclear build can be highly economic against alternatives (principally coal and gas) in many circumstances. Indeed, the upsurge in world gas prices has caused the IEA to completely revise its view of the role of gas in future energy markets – it is now seen as attractive mainly for its assumed quick return on capital rather than fundamentally sound long-run economics. The risks any investor runs with the gas price are certainly emphasised. Coal is, however, making a comeback as a serious baseload generation option, even if it has to bear environmental penalties, such as paying the cost of carbon sequestration or an imposed carbon tax.

Nevertheless, the report doesn't make enough of the superb economics of existing nuclear plants and also doesn't sufficiently question why these cannot be translated into even more new nuclear plants. Some of the relevant factors, principally the various risks involved in new nuclear build, are discussed but in a rather negative fashion. The question, as always, should be: how can a small and environmentally-friendly European country like Finland be building its fifth nuclear reactor while others are either sitting on their hands or even shutting plants down? The IEA will say that this is due to different government policies, but the Finnish investors are looking at prospective costs well below the 5 cents per kWh that the IEA believes is generally achievable for nuclear. This example can no doubt be replicated many times over in many places around the world – it will just take similar drive and initiative from interested local people. It also rather demonstrates one of the weaknesses of the IEA – it's a governmental body and naïvely seems to believe that governments change the world, whereas in fact most evidence suggests that they are usually short-sighted and ineffectual. For example, the strong upsurge in world gas prices was caused by a set of circumstances which were possibly predictable, but in which government policies played only a minor part.

It is also important to note that although both WEO 2006 scenarios show a rather better future for nuclear than they have in the past, in both cases the share of nuclear in world electricity falls from the current 16%. World nuclear generating capacity rises from 364 GWe today to only 416 GWe in the RS by 2030 (representing only 10% of world electricity) and to 519 GWe in the APS (or 14% of the world total). Indeed, the WEO 2006 APS is quite close in overall terms to the WNA reference scenario at 2030, while the RS is between the WNA reference and lower cases.

Upon closer examination the WEO 2006 scenarios seem illogical and inconsistent. For example, in the APS, the OECD North America region shows an increase in nuclear generating capacity from 112 GWe in 2004 to 144 GWe in 2030, while at the same time OECD Europe suffers a decline in nuclear capacity from 132 GWe to 110 GWe. This clearly results from the IEA's extrapolation of changing government policies, but is ridiculous. Given that nuclear is fundamentally more economic in Europe than North America (the former has inferior access to cheap coal), it has governments much more disposed to measures to curb greenhouse gas emissions and suffers from the greater concerns on energy security of supply, a continued slow phaseout in Europe while there is substantial new build in North America looks highly unlikely. And on grounds of consistency, will developing countries be increasing their nuclear capacity from 18 GWe in 2004 to 93 GWe, while Germany is still shutting down serviceable reactors (only one is left in 2030 in the APS) and the UK is not replacing its aged AGRs as they close (there is no new build in the UK, so like Germany, there will be only be one reactor in service in 2030)? China and India certainly have big plans for nuclear, but it seems highly unlikely that Europe will continue to effectively write off this technology option while the remainder of the world is embracing it.

The IEA is wisely not a believer in the 'peak oil' concept, recently put about by many commentators. This is rather like those claiming that uranium could be in short supply by 2030 – clearly both resources are finite, but there should be sufficient available to prevent shortages and huge price swings. The consequence is that the world continues to be fuelled by fossil fuels in both RS and APS, and carbon emissions rise. The APS is successful, however, in making a substantial dent in greenhouse gas emissions by 2030. Despite falling as a share of world electricity, nuclear has an important part to play in this, accounting for about 10% of the reduction from the RS. The majority, however – some 80% – is attributable to energy saving by consumers and also by more fuel-efficient cars and the like.

To stabilise emissions at their 2004 level to 2030 will require a much greater effort. WEO 2006 rather briefly poses a 'beyond the alternative policy scenario' (BAPS), which contains a significantly greater number of conservation measures and also much more clean generation technology – a lot more nuclear and renewables – plus a much higher level of carbon sequestration. Nuclear generating capacity in the BAPS would be 680 GWe in 2030, approaching the level of the WNA upper scenario (740 GWe). The latter, however, is based on the assumption that electricity demand will remain essentially unchanged, so the nuclear share of world electricity will be similar in both (around 20%).

Whatever the quirks of the scenarios in WEO 2006 there is an important final message, which has been well understood by commentators. This is that nuclear needs supportive governments, as it cannot flourish without. This means that in liberalised power markets, where governments will not provide any financial assistance to any generation technology, they still need to have a vision for the national energy mix. They also need to have in place a firm but fair regulatory regime, policies on nuclear waste management and decommissioning and act to assure the public on nuclear proliferation and plant security. The period of energy turbulence since the turn of the new century confirms that electricity cannot be treated like just any good or service in a market environment, as some of the theorists postulated back in the 1980s in encouraging liberalisation. Energy is too important to modern civilisation to allow this and continues to require some special attention from governments, perhaps analogous to the provision of clean water.

The nuclear industry: how strong is it?

Those opposed to nuclear power often make reference to a very powerful 'nuclear industry' which has awesome powers at its disposal to promote its case. In reality, we can see that nuclear is rather different from other energy technologies and the structure of the industry and its supporting professional and trade associations means that it is not quite the monster that is popularly depicted. Indeed, those fighting the case for nuclear often feel that they are fighting with at least one hand tied behind their backs.

Firstly, nuclear lacks a critical mass of strong, powerful companies with good public images to stand up for it. In the oil and gas sector, there are several huge companies such as Shell, BP and Exxon Mobil that devote considerable resources to maintaining a high public profile. Although the activities they engage in may be environmentally unfriendly and unsustainable in the longer term, the recognition and tacit approval they receive as strong corporate beasts means that this tends to get ignored or generally forgotten. With nuclear power, all the revenue comes from the sales of electricity by the companies running the 439 commercial reactors around the world. Although most of these are very large organisations with high public profiles, they generally have little firm commitment to nuclear as they are multi-fuel generators, with oil, coal and gas-fired plants as well – and often also some hydro and renewables too. In some cases they appear to indicate that nuclear technology was foisted on them in the past by public policies and they have had to bear the brunt of problems ever since. Most will speak up in favour of nuclear, saying how important it is for the world's energy future, but without showing any inclination to invest in new plants to make sure that this actually happens.

On the supply side of the nuclear fuel cycle, the position is little better. Only Areva stands out as a substantial international company, with interests in many different areas and countries. As more of its shares gradually become marketed to the general public, this should give a major boost to the nuclear sector as more financial analysts and other commentators will focus on the various activities. The marketing campaign it has adopted over the past few years appears to have been very successful – by first achieving recognition of the Areva name and only then beginning to associate it with the range of fuel cycle activities, which then can be explained to the audience. Although consolidation has taken place within the separate areas of the fuel cycle, such as uranium mining and fuel fabrication, the companies which are strong in particular areas, such as Cameco and USEC are still small compared with the big boys involved in coal, oil and gas. The contracting companies such as Bechtel and Fluor are not attached to any particular technology; similarly the vendors of key plant components such as GE, Alstom and Doosan. There are also a host of consulting companies who find work analysing and advising companies involved in nuclear, but those totally committed tend to be small.

This gives financial institutions and their investors that may believe strongly in the future of nuclear a significant problem. There are very few companies they can invest in to give them exposure, unlike the case if they are bullish about coal, oil or gas. The recent strong upsurge in interest in uranium has brought forward a host of very small junior companies, which are inherently risky as investments. Cameco used to be the only major stock that players can invest in directly, but at least Paladin and Uranium One have emerged as contenders too. Other giants, such as Rio Tinto and BHP Billiton, are involved in many other metals too.

The nuclear industry as it stands is therefore much more fragmented than other energy sectors. Yet this is nothing compared with the various societies, institutes and associations supporting the industry and its people. Despite the huge amount of international trade and specialisation that supports the industry, these are still very much organised at the national level. There is some basis for this as energy policy is still very much a national decision and regional or fully multinational policies are still in their infancy, but it hardly helps to establish a strong and overwhelming case for nuclear within the world's energy future.

In most major countries, there are two types of organisation servicing the nuclear sector. The first are professional societies of nuclear engineers and scientists, such as the American Nuclear Society (ANS), the UK's Nuclear Institute (NI) and the Canadian Nuclear Society (CNS). These have large numbers of individual members who work in companies throughout the nuclear fuel cycle and in universities and research institutes. There are also nuclear branches and divisions of general engineering associations, such as the American Society of Mechanical Engineers (ASME) and the UK Institution of Mechanical Engineers. Secondly, there are organisations devoted to lobbying on behalf of the industry, such as the Nuclear Energy Institute (NEI) in Washington, the Nuclear Industry Association (NIA) in London and the Canadian Nuclear Association

(CNA) in Toronto. These are supported by local companies and concentrate mainly on influencing at the national and local level. In some countries, the professional and lobbying organisations are combined, for example in France with SFEN and in Brazil with ABEN. There is a degree of international cooperation with both professional and lobbying organisations, such as the societies in the Pacific region combining every two years for a Pacific Basin Nuclear Conference, while the larger societies attract participation from overseas at their meetings. In Europe, FORATOM lobbies the institutions in Brussels and Strasbourg on behalf of all the European industry, but the European Nuclear Society, which formerly brought the individual national societies together, is now much weaker.

At the international level, the industry has the World Nuclear Association (WNA) and the World Association of Nuclear Operators (WANO). WNA was formed from the old Uranium Institute in 2001 and runs working groups of members to advance industry knowledge and positions whilst engaging with the international governmental organisations in the UN and OECD systems on behalf of the industry. WANO was formed in the aftermath of the Chernobyl disaster and concentrates on advancing nuclear safety with every plant throughout the world.

While all of these organisations do good work on behalf of the industry, the overall structure is somewhat fragmented and is certainly far from optimal in both servicing the industry's needs for good meetings, better information-sharing to conduct its business and successful lobbying on its behalf. The crazy structure of industry conferences and meetings exemplifies this. Each organisation organises its own events and in some cases relies on these as significant revenue raisers. Although there is a degree of cooperation with respect to timing and coverage, this has proved insufficient to prevent significant overlaps and timing clashes. Industry executives complain that there are far too many conferences that they and their staff should attend, taking up too much travelling time and expense. Alternatively, it has not been unknown for major conferences such as ICONE (the International Conference on Nuclear Engineering) and ICAPP (the International Congress on Advances in Nuclear Power Plants) to occur on the same dates. This shows the need for at least some consolidation and better planning. As there are also nuclear conferences promoted by private organisations such as *Platts* and IBC at irregular intervals, plus other relevant events promoted by UN organisations such as the International Atomic Energy Agency (IAEA), the United Nations Framework Convention on

Climate Change (UNFCCC) and the Intergovernmental Panel on Climate Change (IPCC), the lack of focus and common sense is even clearer.

The first stage is to recognise this serous industry weakness and to address it. It is clear that change is underway in professional and lobbying organisations for a number of reasons. Companies are now more tightly manned and are less willing to give time to employees to engage with their professional societies. An ageing nuclear workforce is also a big factor, together with less willingness of today's young professionals to carry on individual memberships far beyond their student days. This is having a major impact on the national nuclear societies and branches of the professional engineering associations. The lobbying organisations have also to cope with industry consolidation and their smaller numbers of active paymasters need to be convinced that the present structure is optimal or at least workable – which it clearly isn't.

One worthwhile consolidation has been the merging of the former WNA Mid-Term Meeting in the spring with NEI's old Fuel Cycle meeting, held at the same time. This has now worked well on five occasions since 2004, in Madrid, San Antonio, Hong Kong, Budapest and Miami. Another route is to increase support to major annual international meetings in more specialist areas, such as nuclear engineering and waste management. The hope is that these will become bigger and better with increased industry support and that other organisations may merge their overlapping events into them. Hopes may be forlorn, however, as many bodies are over-defensive about their events, failing to recognise the weakening and ageing attendances and are sometimes frightened to reach out beyond their national boundaries.

It is clear that today's structure of companies and representative organisations within the nuclear industry is the result of a complex history. Indeed a strange accident of history. Yet it needs to change if nuclear is to constitute a 'proper' industry and one as strong and powerful as our opponents contend (and we sometimes pretend). The industry itself, on both generation and supply sides, needs to carry on consolidating into a smaller number of more powerful international companies, which can minimise unit costs by engaging in international trade that is as free as practicable. In turn, the other organisations that support it need to be ready to change too and must not get left behind. They should be ready to look in an adventurous way at how they currently conduct their business, to see how both they and their members can prosper rather better.

THE BIG PICTURE

Nuclear: an odd or normal business?

One good approach in giving presentations on nuclear power to outside audiences, is to try to remove some of the emotion that surrounds the business by attempting to demonstrate that it isn't fundamentally 'special' and shares many common features with other sectors, particularly the other energy sources.

As a case in point, although it is possible to draw comparisons with the markets for other metals and minerals, uranium has always been regarded as a rather special commodity. Seen as a key strategic material for military reasons and for energy independence, it has experienced a large amount of government involvement in its production, trade and use. A good contention, however, is that uranium is just another metal commodity, all of which have some special features but also much in common with each other in terms of markets, investor interest and pricing. Uranium's price history is certainly peculiar (with prices lying at a depressed level for many years), but can be explained by special factors, notably the abundance of various secondary supplies. On the question of sustainability of resources, it can be argued that uranium is no different from other commodities whose measured reserves are continuously replenished by changing market prices, providing incentives to new exploration (and eventually new discoveries and later production).

Much of the additional attention that is given to uranium and nuclear is therefore surely quite illogical and unwarranted, imposing substantial costs on the industry, which have to be recovered from customers (effectively all the buyers of electricity). We should therefore try to overcome this, as much as it is possible to do so.

But is this being over-idealistic? Should we just not accept the critical world as it is today, which seems to regard nuclear as so very special? Are we wasting our time in trying to change this perception? Perhaps, as it's a difficult task, but what are the notable features of the nuclear fuel cycle and how special are they?

For one thing, compared with other energy sectors such as oil, coal and gas, the nuclear fuel cycle itself is rather complex. The intermediate stages of conversion, enrichment and fuel fabrication are services provided by specialist companies, while there are important and valuable possibilities for the recycling of materials. The other energy sectors are, by comparison, quite simple in structure. Oil refining, where the barrel of crude oil is split into various products, is a complex process, but not on the same scale as the nuclear fuel cycle.

Compared with the other fuels consumed in generating electricity, the fuel cost in nuclear power is relatively minor compared to total nuclear costs. This remains true even when conversion, enrichment and fuel fabrication costs are added to those of uranium mining, together with an appropriate allowance for the cost of spent fuel management and final waste disposal. Total fuel costs of nuclear are usually under 20% of the total, compared with up to 80% in fossil fuel plants.

Another complexity is that the contractual arrangements normally used within the nuclear fuel market are a peculiarity when compared with trading in other energy commodities. In general, electricity utilities or their procurement agencies contract directly with uranium mining companies for the supply of uranium concentrates. They then have this uranium processed into a useable form through agreements with fuel cycle service providers. Secondary markets for uranium, conversion and enrichment services have, however, also developed and the arrival of ex-military fissile material on the market has added to this.

There is also only a relatively loose short-term quantitative relationship between the annual consumption of nuclear fuel in reactors (*i.e.* reactor requirements) and the utilities' annual demand on uranium producers (*i.e.* procurements). This relationship is complex and is the subject of much analysis and is not generally replicated in other commodities, where production cycles are not so long or inventories so important. Trading in uranium is also distinctive. Most uranium continues to be sold on the basis of multi-annual contracts, based on perceived utility requirements. The spot market in uranium is driven by shorter-term adjustments to utility procurements and by uranium production plans rather than annual reactor requirements. The spot market mainly exists through various traders and brokers. Unlike for many other commodities, there is no terminal clearing market place such as the London Metal Exchange or its equivalents.

Historical uranium production remains highly relevant to the market today. Any nuclear material still containing fissile isotopes can potentially be processed for re-entry into the nuclear fuel cycle. This material may be in the form of depleted uranium from enrichment, reprocessed material from spent fuel, unprocessed spent fuel, or fissile material of military origin. The economics and frequently also the politics of recycling are the limiting factors. Cumulative uranium production throughout the nuclear era, in the period since 1945, therefore retains great importance. But in this respect, the market bears some basic similarities to those of precious commodities such as gold.

There are significant political pressures to reduce the large quantities of surplus ex-military highly enriched uranium (HEU) and military plutonium by using them as fuel within civil nuclear power reactors. In fact, such use of HEU presents few technical difficulties and has already become a major source of supply. This will continue through 2013 and may possibly be extended thereafter. The utilisation of ex-military plutonium in civil reactors, replacing relatively small amounts of fresh uranium, is also now beginning.

Another important feature of the nuclear fuel cycle is its international dimension. Uranium is relatively abundant throughout the Earth's crust, but distinct trade specialisation has occurred, due partly to the low volumes required and therefore the low costs of transportation. For example, uranium mined in Australia can be converted in Canada, enriched in the United Kingdom, then fabricated as fuel in Sweden for a German reactor. Recycled reactor fuel may follow similar international routes, with their related political as well as economic implications. This international dimension, and the existence of large inventories, has also led to the development of an exchanges (swaps), loans and borrowings segment within the nuclear fuel market. On the other hand, there are also trade restrictions which influence the market.

These features above are definitely rather distinctive, but are surely not sufficient to justify the web of licensing, surveillance and national and multinational regulations that are in place throughout the fuel cycle. Added to these impositions, political influence on the nuclear fuel market has also always been a significant factor, with decisions taken to build new reactors, or to allow new fuel cycle facility construction or trade in materials to take place, often containing significant non-economic dimensions.

There must certainly be something else that is driving all the attention given to nuclear. The answer undoubtedly lies in the general fear of radiation and also of nuclear weapons proliferation. This is where the nuclear business is more rationally regarded as rather special and where it is hard to overcome the demands for very special attention. To ensure that safety and non-proliferation objectives are met, a mountain of rules and regulations are administered by governments, regional organisations, such as Euratom in the European Union, and by the International Atomic Energy Agency (IAEA).

The industry can point to its superb record on plant safety (certainly by comparison with the other energy sectors – hundreds of coal mining deaths in China gain little media attention). Weapons proliferation has recently become a bigger issue again, with the concerns over North Korea, Iran and other states, but the Treaty on the Non-Proliferation of Nuclear Weapons (NPT) has undoubtedly been a big success to date – the expectation that many nations would have 'gone nuclear' by now has not been fulfilled.

The difficulty is in analysing how much of these successes have been down to the close attention of the authorities and how much would have happened anyway, without the heavy costs imposed on the industry in compliance. This is very relevant in the United States today, where the industry is putting in a significant effort to streamline the process of getting new reactors built, through initial design certification, early site permitting, and combined construction and operating licensing. Heavy-handed regulation is certainly not appropriate, but the public has the reasonable expectation that the industry will be policed in a way that ensures safety and security goals.

Obviously there is a fine balance to be established here. Getting the favourable messages on industry performance over to both the authorities and the general public is challenging. In particular, the more irrational fears of radiation and the barriers between the civil and military sides of nuclear are hard to explain. Yet the industry and its supporters must never give up here. To expect perceptions and decisions to be wholly rational of course flies in the face of everyone's experience of life, but the facts should be stated calmly and precisely. Attempting to show that the nuclear business is not so odd should form part of this, but we need also not be so defensive about the usual negatives picked on by our opponents. For example, demonstrating the idea that used nuclear fuel is an asset to the industry rather than the industry's Achilles' heel may fly in the face of usual public perceptions. Yet it is a reasonable argument to utilise now. More aggressively promoting it may eventually begin to resonate with key target audiences.

Nuclear: is there any net energy addition?

Whether or not nuclear power plants are built and whether they keep operating for many years after commencing operation is these days essentially an economic decision. The financial costs of construction and operation are compared with the revenues which will flow from selling the electricity generated and a decision can then be made. There are usually alternative options for power available which can be taken up and these can be assessed in a similar way before a generation mix is established.

There are, however, alternatives to financial measures in assessing different modes of electricity generation. One is to use energy itself as a unit of accounting and to attempt to measure the balance between inputs and outputs in the power production process. This is a major element of what is now known as lifecycle analysis (LCA) for generating plants. Such assessments have been made for many years and some have concluded that the nuclear fuel cycle requires very heavy energy inputs to operate, even to the extent that the net energy production will be very low or, in extreme cases, even negative. Such a conclusion would mean that nuclear energy is not sustainable and should therefore not form part of future world energy supply. But how are such conclusions reached?

The first point to make is that analysing energy balances is a complex process because the inputs are so diverse and there is always a question of how far one should go back. It is clear that the energy used in enriching uranium should be included but what about the energy required to construct the enrichment plant in the first place? Clearly we must include the energy requirement for waste management and plant decommissioning. Also the energy it takes to transport materials from A to B? This may be relatively low in the case of nuclear, but coal transportation takes up a huge amount of energy while even natural gas transportation by pipeline is also surprisingly energy-intensive. There are clearly going to be some measurement difficulties – it is easy to measure the energy utilised in operating an enrichment plant but much harder to estimate what is bound up in plant construction or in transporting gas. Finally, it is necessary to establish comparable measuring rods, so kilowatt-hours have to be converted into kilojoules and vice versa, which requires assumptions about thermal efficiencies and the like.

Where nuclear has come out badly in energy balance studies, it is invariably the assessments made for uranium mining and enrichment which are at the source. Such studies have been very pessimistic about the magnitude and quality of uranium resources, such that it is necessary to exploit low grade and inaccessible deposits in the near future, which will involve a greater energy input. In fact, over time, higher grade resources have been discovered and these (particularly in Canada) are the foundation of today's industry. Additionally, in situ leaching (ISL) has become a common technique for exploiting certain low grade uranium resources and this involves a relatively low energy input.

It is, however, the uranium enrichment stage which can potentially involve a very high energy input. Gas diffusion enrichment, the dominant technology until Urenco and the Russians perfected gas centrifuges in the 1980s, uses a huge amount of electricity. If this is supplied by fossil fuel plants, the energy input is very significant, as was the case with the three huge enrichment facilities in the United States – Oak Ridge, Paducah and Portsmouth. It can amount to more than half the lifetime energy input into the fuel cycle. This can, however, be reduced significantly if the electricity is provided by a nuclear power plant, as is the case at the largest gas diffusion plant in the world, Georges Besse in France. Centrifuge enrichment, however, is very economical in energy terms and only uses about 2% of that consumed by a gas diffusion facility. In this case, the total energy input into the entire nuclear fuel cycle will only be about one third of what it will be if gas diffusion enrichment is used.

Even with gas diffusion enrichment, it is clear that the studies which attempt to show that there is little or no net energy gain from nuclear are absurd, relying on unrealistic assumptions about key elements of the fuel cycle. In fact, best estimates of the energy inputs in each area, backed up by a thorough study from Vattenfall of the Forsmark plant, show that the energy inputs in nuclear are at most only 5-10% of the output. Only hydropower can beat this, with both coal and gas lagging well behind. Waste management within the nuclear fuel cycle involves very little use of energy, either in spent fuel ponds, dry cask storage or repositories, although construction of the latter will be relatively energy-intensive.

Renewables are not necessarily so favourable in energy input-output terms. Some studies have even showed wind to be less efficient than nuclear, because of the energy bound up in the steel and concrete for the rotors and the low capacity factors. Similarly, the production of pure silicon for solar photovoltaics requires large energy inputs and accounts for most resource consumption in solar cell manufacture.

The other important area in LCA, in addition to the measurement of energy inputs and outputs, is the assessment of the external costs of power generation, which are the environmental and health consequences which do not appear in the financial accounts. An emerging issue today is the likely contribution of each power generation technology to global warming. It is clear that nuclear does very well on this measure, along with hydro and renewables, as the fossil fuels all emit significant quantities of greenhouse gases. The impact of global warming is, however, still hard to quantify and so some studies ignore this factor.

The ExternE study (2001), launched by the European Union (EU), attempted to provide an expert assessment of lifecycle external costs for European electricity generation. These external costs are those incurred in relation to health and the environment which are quantifiable but are not built into the cost paid by the customer (and so are borne by society as a whole). In particular they include the effects of air pollution on human health, crop yields and buildings as well as occupational disease and accidents. The impact of global warming was excluded, on the basis that it is not yet quantifiable.

The report shows that nuclear incurs only about one tenth of the external costs of coal. This is because the waste costs of nuclear are already internalised, which has the effect of reducing the competitiveness of nuclear when only internal costs are considered, as in conventional financial analyses. The average cost of electricity throughout the EU is 4 cents per kWh without external costs. The externalities of nuclear would add only 0.4 cents per kWh to this, whereas coal's add more than 4 cents per kWh (*i.e.* the external costs are above the internal cost), gas's add 1.3-2.3 cents per kWh. Only wind shows up better than nuclear, adding only 0.1-0.2 cents per kWh. So if the external costs could be incorporated, the cost of coal-fired electricity would double and gas-fired would rise by around 50%. If the potential impact of fossil fuels on global warming was to be added, the impact would be considerably greater.

Economists would argue that these external costs should be incorporated in the electricity price paid by consumers or else there is a misallocation of resources. This could ideally be achieved by imposing appropriate taxes to reflect the external costs with the revenues from these sufficient to compensate society. This would considerably alter the mix of electricity generating capacity in favour of those with low external costs, essentially nuclear, hydro and the renewables. However, there are substantial political barriers to this being achieved as society has always effectively subsidised fossil fuel use, as pollution costs have been ignored.

Nuclear power therefore comes out very favourably from LCA, whether this is on the basis of merely looking at energy inputs and outputs or also incorporating external costs. The analyses which purport to show the opposite can easily be shown to be misguided. The only possible cloud on the nuclear horizon is used fuel management. Critics would argue (with some justification) that until further management decisions have been made and repositories have entered operation, the long-term costs of various solutions are hard to calculate. The industry can counter by producing various estimates, but it will take time for the true picture to emerge.

THE BIG PICTURE

Nuclear proliferation and the terrorist threat:
a barrier to new build?

With all the attention in the media granted to North Korea's October 2006 nuclear test and Iran's alleged intentions to pursue a weapons programme, the fear that nuclear proliferation may cast a shadow over the nuclear renaissance has emerged. Critics (for example see Mycle Schneider's report for the European Parliament Greens, *The Permanent Nth Country Experiment – Nuclear Weapons Proliferation in a Rapidly Changing World*) allege that nuclear energy and bombs are merely two faces of the same coin. Nuclear materials could conceivably be diverted from a civil nuclear power programme into the production of nuclear weapons or alternatively, major fuel cycle processes (notably enrichment and reprocessing of used fuel) could be employed to produce weapons, rather than fuel for civil reactors. A related concern is over security of civil nuclear facilities, which has multiplied since the 9-11 terrorist attacks in New York and Washington. The possibility of aircraft crashing into such plants has been raised, as has terrorist incursions at plants either to acquire materials for weapons or to misuse the facility to create an explosion or a major radioactive release.

Security has been addressed by deploying additional armed personnel at facilities and by other measures to prevent incursions, while new nuclear plants are designed with the possibility of an aircraft impact much in mind. Although such events are clearly not impossible, the entire 50-year history of civil nuclear power contains nothing to suggest that the risks are other than very remote. Little can be done other than what has been accomplished already and the risks should certainly not be allowed to determine future actions. To keep things in perspective, should London's new Wembley Stadium not be licensed for 80,000 football fans, simply because a direct aircraft strike during a game could conceivably kill thousands?

Proliferation of materials and technology misuse are clearly more substantive risks, particularly as it will likely involve sovereign states with their greater resources above those of a terrorist organisation. Critics of nuclear power emphasise that designing a nuclear bomb is not particularly difficult. This, in itself, doesn't create a great risk if the necessary plutonium or highly enriched uranium is not available either by diversion or production in a local facility. It is therefore necessary for the anti-nuclear forces to focus on alleged weaknesses in the international nuclear safeguards regime, stories of illicit materials trafficking, supposed limitations in security of nuclear materials transport and the possible spread of enrichment and reprocessing technologies to countries that may have an interest beyond normal civil uses.

Probably the greatest weakness of the antis case is that it hasn't happened yet, despite considerably slacker arrangements in the past than are present today. While there is no room for complacency and further strengthening of arrangements is fully warranted, the risks are, in reality, actually as remote as those associated with plant security.

Over the past 35 years the International Atomic Energy Agency's (IAEA's) safeguards system under the Treaty on the Non-Proliferation of Nuclear Weapons (NPT) has been a conspicuous international success in curbing the diversion of civil uranium into military uses. It has involved cooperation in developing nuclear energy while ensuring that civil uranium, plutonium and associated plants are used only for peaceful purposes and do not contribute in any way to proliferation or nuclear weapons programmes. In 1995 the NPT was extended indefinitely.

Most countries have renounced nuclear weapons, recognising that possession of them would threaten rather than enhance national security. They have therefore embraced the NPT as a public commitment to use nuclear materials and technology only for peaceful purposes. The NPT's main objectives are to stop the further spread of nuclear weapons, to provide security for non-nuclear weapon states, which have given up the nuclear option, to encourage international cooperation in the peaceful uses of nuclear energy, and to pursue negotiations in good faith towards nuclear disarmament leading to the eventual elimination of nuclear weapons. It is clearly the last objective where least progress has been made, as the five weapons states (the USA, Russia, China, France and the UK have arguably failed to keep to their side of the bargain.

The IAEA undertakes regular inspections of civil nuclear facilities to verify the accuracy of documentation supplied to it. The agency checks inventories and undertakes sampling and analysis of materials. Safeguards are complemented by controls on the export of sensitive technology from countries such as the UK and the USA through voluntary bodies such as the Nuclear Suppliers Group (NSG).

Parties to the NPT agree to accept technical safeguards measures applied by the IAEA. These require that operators of nuclear facilities maintain and declare detailed accounting records of all movements and transactions involving nuclear material. Over 550 facilities and several hundred other locations are subject to regular inspection and their records and nuclear material being audited. Inspections by the IAEA are complemented by other measures such as surveillance cameras and instrumentation.

The aim of traditional IAEA safeguards is to deter the diversion of nuclear material from peaceful use by maximising the risk of early detection. At a broader level they provide assurance to the international community that countries are honouring their treaty commitments to use nuclear materials and facilities exclusively for peaceful purposes. In this way safeguards are a service both to the international community and to individual states, who recognise that it is in their own interest to demonstrate compliance with these commitments. All NPT non-weapons states must accept these full-scope safeguards. In the five weapons states plus the non-NPT states (India, Pakistan and Israel), facility-specific safeguards apply. IAEA inspectors regularly visit these facilities to verify completeness and accuracy of records.

The terms of the NPT cannot be enforced by the IAEA itself, nor can nations be forced to sign the treaty. In reality, as shown in Iran and North Korea, safeguards can be backed up by diplomatic, political and economic measures. Iran and North Korea illustrate both the strengths and weaknesses of international safeguards. While accepting safeguards at declared facilities, Iran set up equipment elsewhere, allegedly in an attempt to enrich uranium to weapons grade. North Korea used research reactors (not commercial electricity-generating reactors) and a reprocessing plant to produce some weapons-grade plutonium. The weakness of the NPT regime lies in the fact that no obvious diversion of material has been involved. The uranium used in the reactors came from either indigenous sources or that supplied purely for research purposes, and the countries themselves built the nuclear facilities concerned, without declaring them or placing them under safeguards arrangements.

In 1993 a programme was initiated to strengthen and extend the classical safeguards system, and a model protocol was agreed by the IAEA Board of Governors in 1997. The measures boosted the IAEA's ability to detect undeclared nuclear activities, including those with no connection to the civil fuel cycle. The so-called Additional Protocol gives the IAEA considerably more information on nuclear and nuclear-related activities, including R&D, production of uranium and thorium (regardless of whether it is traded) and nuclear-related imports and exports. Inspectors also have greater rights of access and will include any suspect location. Visits can be at short notice (*e.g.* two hours), and the IAEA can deploy environmental sampling and remote monitoring techniques to detect illicit activities. As of the beginning of 2008, of the order of 130 countries had Additional Protocols either in force or approved and signed.

The greatest risk of nuclear weapons proliferation has traditionally rested with countries which have not joined the NPT and which have significant unsafeguarded nuclear activities. India, Pakistan and Israel are in this category. While safeguards apply to some of their activities, others remain beyond scrutiny.

A further concern is that countries may develop various sensitive nuclear fuel cycle facilities and research reactors under full safeguards and then subsequently opt out of the NPT. This suggests that moving to some kind of intrinsic proliferation resistance in the fuel cycle is timely. There are a number of ideas, previously floated many years ago, which have been dug out and revamped. One key principle is that the assurance of non-proliferation must be linked with assurance of supply and services within the nuclear fuel cycle to any country embracing nuclear power. The Global Nuclear Energy Partnership (GNEP) programme announced by the United States and complementary initiatives discussed by Russia and the IAEA might guarantee the supply of nuclear fuel and services for *bona fide* uses, thereby removing the incentive for countries to develop indigenous fuel cycle capabilities. Creating new multinational, possibly regional, fuel cycle facilities for enrichment, reprocessing and used fuel management, based on joint ownership is one concept, as is reinforcing existing commercial market mechanisms of long-term fuel supply contracts, possibly involving fuel leasing and the take-back of used fuel, so obviating the need for fuel cycle facilities in most countries. Yet there is clearly a risk here of dividing the world into 'good guys' and 'bad guys', in a politically discriminatory way.

One stimulus to rejigging old rules may be the new arrangements on nuclear trade the United States is now finalising with India, which has been many years out in the cold owing to its weapons programme. There remain substantial challenges in implementing this, particularly with the NSG arrangements, but something needed to be done, as categorising the second most populous nation in the world as a 'nuclear outlaw' was never helpful and, if anything, hardened attitudes.

To summarise, a lot remains to be done to strengthen non-proliferation arrangements, but progress is slowly being made. Possible diversion of fissile materials to illicit uses is likely to come up as an issue when any new nuclear build programme is proposed, but the risks are, in reality, low and tolerable. It is unclear why building lots of new nuclear power plants will markedly increase any of these risks, particularly if they're in countries where nuclear is already well-established. The number of new countries likely to get nuclear plants by 2020 is in any case quite small and these will be expected to embrace best international practice. As a parallel, the small risks involved in international air transport do not prevent passengers from cheerfully undertaking many flights, as the risks are regarded as low and well-managed. Nuclear power must be seen to be doing the same.

GNEP: the right way forward?

The Global Nuclear Energy Partnership (GNEP) proposed by US President Bush in February 2006 has received a great deal of publicity, much of it centred on the apparent change in US strategy towards reprocessing used fuel, rather than carrying on with the policy of final disposal in repositories. There is, however, much more to the initiative than this and it is worthwhile examining the obvious ways in which it addresses many of the awkward issues currently faced by the nuclear industry. But at the same time, there are undoubtedly some serious difficulties on the road ahead to ensuring the concepts become reality.

There is some debate as to where the real motivation for GNEP lies in the US Administration as its announcement seemed somewhat sudden and there are accusations (in common with the initiative on nuclear trade with India) that not enough consultation, both within and outside the US, had taken place first. It is possible, however, to see a gradual move in thinking within the US since about 2001 towards the advantages of reprocessing and GNEP merely makes this explicit. For several years, there has been interest in new forms of reprocessing which do not separate plutonium from uranium (in fact recovering both together), and which segregate other actinides from fission products, enabling the actinides to be burned. Indeed, the US budget for 2006 included $50 million to develop a plan for 'integrated spent fuel recycling facilities'. Nevertheless, the GNEP announcement can certainly be depicted as a major shift in official US policy, which has been wedded to the 'once through' fuel cycle since the Carter Administration in the late 1970s.

The two most important areas addressed by GNEP are the concerns about proliferation of nuclear weapons and the difficulties the industry continues to experience in developing coherent policies for used fuel management.

The challenges of new countries utilising nuclear technology and the variability of political will when confronted with situations such as Iran suggest that moving to some kinds of intrinsic proliferation resistance in the fuel cycle is timely. One key principle is that the assurance of non-proliferation must be linked with assurance of supply and services within the nuclear fuel cycle to any country embracing nuclear power. Impetus had already been given to this by Mohammed ElBaradei, Director General of the International Atomic Energy Agency (IAEA), who pointed to the need for better control of both uranium enrichment and plutonium separation at the UN General Assembly in October 2005.

There remains the issue of who runs these multilateral initiatives – certainly the US or Russia can be the inspiration but it seems preferable for the process to be under IAEA control or coordination. It is clear that GNEP must work within existing international arrangements. There are already several approaches under discussion by an expert group convened by the IAEA, including: developing and implementing international supply guarantees with IAEA participation, for example with the IAEA as administrator of a fuel bank; promoting voluntary conversion of existing facilities into multinational control, including the non-signatory countries to the NPT (such as India and Pakistan); and creating new multinational, possibly regional, fuel cycle facilities for enrichment, reprocessing and used fuel management, based on joint ownership. A further idea is to reinforce existing commercial market mechanisms of long-term fuel supply contracts, possibly involving fuel leasing and the take-back of used fuel, so obviating the need for fuel cycle facilities in most countries.

The other important matter addressed by GNEP is used fuel management. A significant part of the incentive of advanced reprocessing technologies is to reduce volumes of high-level wastes and simplify their disposal. This does not, however, mean that waste repositories such as Yucca Mountain will never be needed – they must still be planned for and developed, but the quantities of material destined for them will be much reduced. The difficulties encountered with establishing Yucca as an operating repository have undoubtedly influenced the move towards GNEP. The likelihood of having to establish several Yuccas in the United States alone, if there is a significant boom in nuclear power in the 21st Century, has obviously concentrated a lot of official thinking. Thinking expansively, Yucca may no longer now be seen as a repository for the used fuel currently in storage at reactor sites throughout the US, but as a facility for the receipt of the final wastes from future reprocessing activities. In other counties too, there also seems to be a shift in attitudes about the value of used fuel that could eventually have repercussions for many national waste management programmes. Some facilities currently envisaged as final disposal repositories may only be used for interim storage of spent fuel that will eventually be reprocessed and recycled, hence the trend to retrievability. But this is running some way ahead – the current plan in the US remains to get Yucca Mountain licensed as a repository for the used fuel as it currently exists, as without this 'final solution' it will be hard to license new reactors in the US.

In moving towards longer-term storage of used fuel with the expectation that it will eventually be reprocessed, it is important to demonstrate that the industry is not just 'passing the buck' to the next generation. Used fuel must begin to be somehow presented as an asset, as a key foundation for fuelling the next generation of reactors, without the need to mine and utilise greater quantities of a finite resource, such as uranium, than is really necessary. Over two million tonnes of uranium have been mined since 1945, both for military and civil nuclear programmes. It makes sense in the future to use as much as possible of what were formerly regarded as wastes from previous nuclear operations as true fuel assets in new reactor types.

This leads to a further important part of GNEP, which is the link with the Generation IV programme and other advanced reactor initiatives. Reactor systems with full actinide recycling as part of a closed fuel cycle will produce very small volumes of fission product wastes without the long-life characteristics of today's used fuel, and will have high proliferation resistance. The 'classic' closed fuel cycle with aqueous (PUREX) reprocessing and recycling of plutonium into mixed oxide (MOX) fuel is not intrinsically proliferation resistant. There are, however, already significant quantities of separated civil plutonium, reprocessed uranium and depleted uranium in inventory and these may well be utilised when the new reactor designs become reality. Although it is almost certain that fresh uranium will still have to be mined, the quantities will be much lower than required by the current generation of reactors. Although the nuclear industry is convinced that there are more than adequate uranium reserves and resources to fuel any conceivable growth path of nuclear energy this century, the higher uranium prices which are likely to be necessary to develop all the new mines will make recycling uranium and plutonium from used fuel relatively more attractive in an economic sense. Despite the recent trend towards higher uranium prices, a significantly lower fresh fuel input may allow the next generation of reactors to be fuelled even more cheaply than those of today. Low and relatively stable fuel prices are already a significant advantage of the current generation of reactors against alternative fossil fuel generating modes, but the future looks even better.

Finally, it can be argued that GNEP makes a contribution to the idea, explicit within the NPT, of the leading nuclear nations spreading the benefits of nuclear technology to other countries. After the provision of many research reactors in the early days, the United States and the other nuclear weapons states have done relatively little in this regard. It may also be argued that the economies of scale in enrichment and reprocessing plants and eventually waste repositories, suggest that there should be only a small number of facilities worldwide. Although developing national facilities may appear to meet some immediate local objectives, in the long run it would be better from the economic standpoint to re-deploy the resources elsewhere and buy, with guarantees, from abroad. The current national repository solutions that are posed certainly make little sense either economically or politically. But moving to an international regime requires substantial changes to the rules of nuclear commerce as they currently stand.

This leads to the first of the difficulties in getting GNEP up and running. Some will argue that it is very ambitious on several counts – the new technology required, the timescales quoted and particularly the wholesale changes to the current international arrangements. Yet these are clearly in need of reform – they have worked rather well in the early days of nuclear power, but if a much more expansive future is foreseen, with thousands of reactors being built to satisfy the world's need for cheap power, potable water and hydrogen, some fundamental reforms are needed. Tinkering with the existing arrangements will not be enough.

There are also some concerns about the extent to which following GNEP will upset existing nuclear research programmes in particular countries and also the current plans for the fuel cycle. There are always strong vested interests in continuing along the same path. In Japan, for example, there are fears that having just re-established its nuclear programme on a twin platform of new reactor construction combined with the reprocessing of used fuel at Rokkasho and subsequent recycling of plutonium in light water reactors, GNEP may prove to be a diversion. Having struggled to obtain public acceptance for reprocessing and subsequent recycling, the implication within GNEP that a much superior reprocessing technology is just around the corner may pose local difficulties. If the GNEP proposals threaten to bring foreign used fuel to Japan for reprocessing, this is politically a very hot potato to handle.

On the other hand, some have claimed that GNEP really doesn't go far enough. If we're going to complete a 'once per century' reform in international nuclear arrangements, GNEP may look too closely at present day proliferation and waste concerns, rather than the challenges of reaching a more expansive nuclear future. Its concentration on 'burner' rather than 'breeder' reactors fits in with the proliferation concerns, but does less to promote the vision of thousands of future reactors. If we're going to have an energy world fuelled largely by hydrogen, we need to create a better link between today and the future, but GNEP only goes a limited distance in this.

One barrier to the creation of multinational fuel cycle facilities, with attendant guarantees of supply in exchange for strict adherence to safeguards, is the view held my some countries in the world that they ought to develop full fuel cycle facilities because of security of supply or import-saving reasons. Transport of nuclear fuels from continent to continent has also become difficult, to add to concerns about the reliability of various suppliers, so there is some argument for developing facilities 'at home'. For example, countries possessing significant uranium resources are inclined to develop them and then think about developing other areas of the fuel cycle too. Hence Brazil's involvement in uranium and enrichment, to fuel its own reactors and, less obviously, the views now regularly expressed in Australia that it should 'add value' to its uranium sales by converting and enriching too. Becoming regional fuel cycle centres under full IAEA safeguards may cover these aspirations as the number and location of these is yet to be specified. The economic case with economies of scale suggests, however, that there should be relatively few large facilities worldwide.

An alternative view of GNEP may see it as somewhat discriminatory and potentially anti-competitive. By restricting parts of the fuel cycle to particular countries, albeit with fair rights of access to nuclear materials, there is a risk of maintaining or even reinforcing the existing NPT arrangements that have always upset certain nations, notably India and Pakistan. Similarly, by maintaining a market stranglehold on, for example, enrichment facilities in the existing countries, it can be argued that the market will be uncompetitive and lead to excessive profits being achieved by those who are so favoured. The more expansive nuclear vision surely needs it to become a more 'normal' business as far as public acceptance is concerned, with as few special provisions and restrictions as possible. So somehow a reasonable balance has to be established, which hits many possibly-competing objectives.

Finally, it is clear that GNEP must overcome a number of difficulties of coordination in both the US itself and also internationally. Nuclear policy within the US often gives the impression of too many initiatives and too little action and it is not surprising that Congress sees fit to allocate or withdraw funding in seemingly inconsistent ways. The overall nuclear programme needs to be made more coherent, with full integration of the plans for new reactors, used fuel management and the more visionary goals of GNEP. The full funding requested for GNEP has unfortunately already been cut, which doesn't bode well. Similarly, on the international stage, GNEP must be integrated with what is happening under IAEA auspices and also the plans announced by President Putin for international fuel cycle facilities located in Russia. Although it is good that several parties are thinking along the same lines, it is necessary for the plans in so important an area to be properly coordinated without national interests holding too much sway.

Changing international arrangements: what are the new initiatives?

The Global Nuclear Energy Partnership (GNEP) continues to attract a good deal of publicity and comment, both positive and negative. The opposition depicts it as a flagrant example of US arrogance, embracing dangerous proliferation-prone technologies whilst claiming to make the world a safer place. Supporters, on the other hand, regard it as a fine example of the United States assuming world leadership in encouraging the world to open up important questions which have been avoided during the 'dark ages' of nuclear power since Three Mile Island and Chernobyl, in order to encourage a new era of booming nuclear commerce and international cooperation.

The truth, of course, lies somewhere between these two extremes. It is important, however, to place GNEP within the wider context of a number of other complementary developments taking place simultaneously. Although there is tendency to pronounce on each of these in isolation, they must be seen as part of a bigger picture. Firstly, there are the nuclear cooperation agreements that the United States has been working on with India, Russia and other countries. The aim of these is to facilitate nuclear commerce and cooperation in order that restrictions can be removed. India is being brought back into the fold, having been isolated from nuclear commerce by its refusal to sign the Treaty on the Non-Proliferation of Nuclear Weapons (NPT) and its subsequent testing of nuclear weapons. This breakthrough doesn't please everybody and leaves open the question of what to do with Israel and Pakistan, also non-NPT signatories and (assumed in the case of Israel) effectively nuclear weapons states. Yet something had to be done as the world's largest democracy could not be left isolated from international non-proliferation arrangements. The solution is certainly imperfect, but the alternative of inaction is much more unpalatable. The bilateral agreement with Russia, which should eventually lead to a similar so-called '123 agreement' as with India, outlines a framework for the global sharing of nuclear expertise and technical assistance. It fits in well with GNEP as it aims to provide modern and proliferation-resistant reactors to third countries and help develop used fuel solutions, so they have incentives to develop nuclear energy safely and without attendant security risks.

Secondly, there are the initiatives stimulated by both the International Atomic Energy Agency (IAEA) and Russia to develop arrangements and facilities that might guarantee the supply of nuclear fuel and services for *bona fide* uses, thereby removing the incentive for countries to develop indigenous fuel cycle capabilities. It is clear that GNEP must work within these other arrangements as there are already several approaches under discussion, including: developing and implementing international supply guarantees with IAEA participation (for example with the IAEA as administrator of a fuel bank); promoting voluntary conversion of existing facilities into multinational control; and creating new multinational, possibly regional, fuel cycle facilities for enrichment, reprocessing and used fuel management, based on joint ownership. The Russians have already opened a dedicated international enrichment centre at Angarsk but it is future used fuel facilities that are likely to be most attractive to other countries.

Thirdly, there are more specific arrangements in nuclear fuel trade, which are subject to reform. Some of these are included in further bilateral nuclear cooperation agreements, such as between Australia and China, allowing Australian suppliers to help satisfy rapidly-growing Chinese demand. The most important, however, is the future of the Suspension Agreement that originally imposed antidumping duties on Russian uranium imports to the United States in 1992. The main impact of this today is to prevent US utilities from contracting directly with the Russians on uranium enrichment services, something they would very much like to do in the context of stimulating greater market competition. An amendment was agreed in early 2008 to allow Russia at least some direct access to the US market, particularly after 2013, when the agreement on downblended highly enriched uranium (HEU) between Russia and the US expires.

Finally there are other developments in the nuclear fuel cycle are relevant to GNEP. In particular, there is clearly a move back towards the idea of supplying fabricated fuel as a package, rather than a utility buying uranium, conversion, enrichment and fabrication services separately. New commercial arrangements amongst the ranks of the reactor vendors/fuel fabricators in the aftermath of Toshiba's acquisition of Westinghouse (particularly the reshuffling of the relationships between Western and Japanese companies) is leading to stronger partnerships. Big uranium suppliers, notably Kazatomprom with its acquisition of a 10% stake in Westinghouse, are seeking vertical integration to add value to their prime asset, while the fabricators are interested in offering a packaged fuel service. This has been generally avoided since Westinghouse burned its fingers very badly in the late 1970s, when it was caught out by rapid uranium price inflation. The Russians, however, have always offered a 'cradle to grave' fuel supply service, even taking back the used fuel for reprocessing and/or disposal. For countries acquiring nuclear reactors for the first time, such packaged fuel services are very attractive but they can also fit in with the non-proliferation objectives of GNEP and the other international initiatives, as they obviate the need to establish domestic facilities.

One question is whether GNEP and these other initiatives constitute sufficient underpinnings for a brave new world of nuclear power or is something else needed? It is clear that many changes need to be made if nuclear is to fulfil its potential as a vital element in a clean energy future. Many of the institutional and commercial arrangements in the industry have remained frozen in time throughout the long period when nuclear was seen to stagnate. New challenges have emerged, notably the increased focus on non-proliferation and plant security concerns, which have to be addressed.

Indeed, the continued delays at the Yucca Mountain waste repository project argue that the 'closed fuel cycle' provisions of GNEP are needed more than ever. Some US government money has been provided to continue work on advanced reprocessing technologies aiming to reduce volumes of high-level wastes and simplify their disposal, while the mixed oxide (MOX) fuel fabrication facility at Savannah River is now under construction. This does not, however, mean that waste repositories will never be needed – they must still be planned for and developed, but the quantities of material destined for them will be much reduced. In other countries too, there also seems to be a shift in attitudes about the value of used fuel that could eventually have repercussions for many national waste management programmes. Some facilities currently envisaged as final disposal repositories may only be used for interim storage of spent fuel that will eventually be reprocessed and recycled, hence the trend to retrievability.

To conclude, it can be said that GNEP and the other initiatives are important elements in putting international nuclear policy on the right road for the future, but are not in themselves sufficient. There are a huge number of initiatives which require coordination in both the US itself and also internationally. Nuclear policy within the US often gives the impression of too many initiatives and too little action and it is not surprising that Congress sees fit to allocate or withdraw funding in seemingly inconsistent ways. The overall US nuclear programme needs to be made more coherent, with full integration of the plans for new reactors, used fuel management and the more visionary goals of GNEP. Then it is necessary for the plans in so important an area to be properly coordinated without national interests holding too much sway. But this is a rare area of international policy where the Bush Administration has got most things mainly right and where a degree of US leadership should be entirely welcomed.

SUMMING UP

The statement 'nuclear power is at a crossroads' has become almost a cliché, and those within the industry are now tired of hearing it. It feels rather like the jury being sent out to consider its verdict at a major trial, only to be unable to reach a verdict. In recent years, the amount of positive news about nuclear has increased enormously, but the future is still far from certain. Although the technology is mature, the issues discussed in this book demonstrate that returning to the rapid growth period of the 1970s and 1980s will not be easy. Ultimately, the hope is that the 1990s and the early years of this century will be seen as an aberration, with the industry pausing for breath while particular issues are resolved, before returning to the earlier growth rates. While great strides have been made in facilitating this, there remains a lingering doubt that nuclear programmes are just too complex for both countries and individual commercial companies to undertake, and that easier (but ultimately inferior) options may be chosen.

The environmental credentials of nuclear are now increasingly accepted, even by many of those previously strongly opposed to its deployment. Global warming is potentially the biggest issue facing mankind this century and nuclear can play a substantial role in its mitigation. This may be somewhat limited at the beginning but will become substantial once new build programmes get underway. For the immediate term, the prime motivation of countries such as China and India moving strongly towards nuclear is more to do with the direct impact of burning coal on urban air pollution, which is killing thousands of their citizens today.

As world oil and gas prices reach record levels, the energy security advantages of nuclear also look increasingly attractive. In the medium term, it offers the possibility of avoiding some of the adverse geopolitical disadvantages of over-dependence on fossil fuels, by direct substitution in electricity generation. Ultimately, however, it is with a complete transformation of the transportation sector in favour of hydrogen fuels that nuclear offers much more, both in terms of energy security and environmental satisfaction. This would involve a major adjustment of the current world energy economy, but in many ways less than that championed by the environmental movement. A strong move towards a small-scale, decentralised energy production system, with significant conservation of resources and renewables taking the load, is indeed superficially an attractive option. Yet it potentially represents an even greater break with recent history.

The economic credentials of nuclear naturally look much more attractive when oil, gas and coal prices are at new, higher levels. It is now proven that nuclear plants can run both safely and highly economically for upwards of 40 years; the bigger issue is getting them started up in the first place. Once in operation, assuming that they are online for 90% of the time, costs of nuclear generation will be very low and highly attractive for all customers. Even with the uncertainties about new build costs today, new nuclear plants should also be attractive financial propositions in many power markets, particularly if carbon is penalised. The risks of nuclear investments are obvious and must be allocated amongst those parties best placed to bear them. But projects can certainly be structured in such a way that this can be accomplished. Where the economic balance is fine, there may be good arguments in favour of some initial governmental support in order to encourage nuclear investment (as is currently the case in the United States), but this should be unnecessary once initial reactors have been completed. Costs of subsequent units should be substantially lower and cost certainty greater than today, when there is naturally concern about how much such large projects will entail.

The best thing the industry can do to gain public acceptance, always flagged up as a vital issue, is merely to carry on doing precisely what it has been doing over the past ten or so years. Operating the existing plants safely and economically while being as open as possible with information about both specific operations and nuclear in general, has proved very effective. Nuclear will always be under intensive scrutiny, but most people don't want to be bothered by energy issues, with their concern not going much beyond receiving a reliable and cheap supply. Carrying on doing this in a quiet but effective manner is the best recipe for the future.

A few new issues brought up against nuclear today are essentially non-issues, but this hasn't been helped by the industry talking so much about them. Some shortages of staff are inevitable as uranium mining picks up once again and will also be experienced as other sectors gear up to renewed growth. But these will rapidly be overcome as people respond to the economic incentives, absent for so many years. The same applies to the supply of key nuclear plant components. It would be amazing if facilities were ready now to fulfil all the possible new orders, but the industry has geared up from a similar low base once before. Certainly the prospect of completing

20-30 new reactors a year in the 2020s is a challenging one, but far from impossible from where we are today. Indeed, a really substantial push towards nuclear could lead to an even higher rate of new reactor building. Think of France with 50 million people opening 3-5 new reactors a year in the 1980s and extrapolate that to China and India alone.

Whether nuclear power can be established in many new countries by 2020 remains somewhat doubtful. The number of likely new countries by then can be counted on the fingers of one hand – perhaps Indonesia, Vietnam, Egypt, Turkey and (perhaps an adventurous choice) Italy can be regarded as in the forefront, but things could change rapidly. Many counties are currently talking about nuclear programmes, but the time lags involved suggest that for most this will mean the 2020s rather than before. It is unlikely, however, that nuclear will be established in many new countries until the mooted renaissance is confirmed in many of the developed countries and the programmes in China and India got rather further ahead. Perhaps Germany can be different and phase out nuclear while similar countries are doing precisely the opposite, but this can't spread too far or else it could prove contagious. Countries can make their own choices in energy matters, but these are not completely independent – orders for new reactors in North America and Western Europe are essential for the industry to step boldly forward.

The time is now ripe for all the fine words about nuclear to be put into practice. The era of decision-makers sitting on the fence and the hung jury has to be over. Getting more and more people onside – governmental, industrial and financial – has gradually been achieved and a consensus is almost reached, in most countries at least. What remains missing is leadership. Both from governments, who must create the right framework, and from the industrial leaders committed to carrying the industry forward. Their vision has to be an expansive one, not just in terms of the number of new reactors but also of a 'new' industry, rather different from what we have today. Standardised reactors with modular construction, produced by a small number of companies and easily licensed in many world markets, have to be the way for the future. The nuclear fuels business also needs to undergo a transformation, throwing off its legacies from the past, in order to serve vibrant new markets. The magnitude of the Chinese reactor building programme in the period to 2020 will in itself fundamentally change the balance of the nuclear business worldwide, but many reactor orders in other countries are essential.

Although technically mature, the industry is arguably still at an early phase in its lifecycle.

What is needed is the drive, initiative and also courage to see through these challenging long-term investment projects. The political cycle is short, as is the usual time in top office for corporate staff, but both groups must see the wisdom of pushing hard for something that is demonstrably worthwhile. The short-term disadvantages, where resources are poured in without much obvious return, have to be placed into their correct context. If they can't, nuclear will not quickly disappear, but will likely fade away, slowly but surely.

GLOSSARY

ABACC	Argentine-Brazilian Agency for the Accounting and Control of Nuclear Materials, a nuclear safeguards organisation
ABWR	Advanced Boiling Water Reactor, developed by General Electric and its Japanese partners
AECL	Atomic Energy of Canada Limited, a publicly-owned Canadian nuclear technology developer and reactor vendor
AHWR	Advanced heavy water reactor, a PHWR being developed in India to run largely on thorium fuel
AP1000	A Generation III PWR developed by Westinghouse
APR-1400	Advanced Power Reactor, a PWR developed by KOPEC from CE technology
APWR	Advanced Pressurized Water Reactor, a large Generation III PWR developed by Mitsubishi
ASE	Atomstroyexport, a Russian reactor technology developer and reactor vendor
BNFL	British Nuclear Fuels Limited, a publicly owned (but now being privatised) nuclear services company
BP	British Petroleum, a major oil and gas company
BWR	Boiling water reactor, one of the two variants of the LWR
CANDU	Canadian Deuterium Uranium reactor, the most common variant of the PHWR
CCGT	Combined cycle gas turbine, the latest generation of gas-fired power station design containing a gas-fired turbine generator and a steam turbine generator employing steam produced from the gas turbine exhaust heat
CDU	Christian Democratic Union, the conservative party in Germany

CE	Combustion Engineering, a reactor vendor acquired by Westinghouse
CEZ	The electricity generating company in the Czech Republic
CNEA	Comisión Nacional de Energía Atómica, the Argentinian governmental nuclear body
COL	Combined construction and operating licence, a streamlined reactor licensing process in the United States
CPR-1000	A Chinese PWR developed from French technology
CSU	Christian Social Union, Bavarian partners of the CDU in Germany
EDF	Electricité de France, a major company generating nuclear electricity
EPR	European Pressurized Water Reactor, a large evolutionary PWR developed by Framatome and Siemens (now Areva NP)
ESP	Early site permit, an initial stage in the streamlined United States licensing system
EU	European Union, a partnership of 27 European countries
FBR	Fast breeder reactor, a fast neutron reactor configured to produce more fissile material than it consumes
FOAKE	First-of-a-kind engineering, referring to additional costs specifically attributable to initial units of a new reactor series
G8	Group of Eight, comprising the leading developed countries of the world
GDP	Gross domestic product, a measure of the economic output of a country or region
Gen III/IV	Generation III and Generation IV, successive generations of reactor development
GIF	Generation IV International Forum, an international collaborative effort on next generation nuclear reactor development

GLOSSARY

GJ/h	Gigajoules per hour, a unit of energy consumption
GNEP	Global Nuclear Energy Partnership, an international nuclear cooperation regime proposed by the United States
GWe	Gigawattt electric, a measure of electricity generating capacity
HEU	Highly enriched uranium, having a uranium-235 assay increased from the natural 0.7% to (usually) over 90%
HLW	High-level waste, having a high level of radioactivity and therefore requiring careful handling, care and disposal
HTGR	High temperature gas cooled reactor, a type that is under development but has not, so far, achieved commercial success
I&C	Instrumentation and control, equipment which monitors and operates an industrial plant
IAEA	International Atomic Energy Agency, the United Nations body designated to deal with nuclear energy matters
IEA	International Energy Agency, the OECD body designated to deal with energy matters
ILW	Intermediate-level waste, having a middle level of radioactivity
INB	Indústrias Nucleares Brasileiras, the state-owned Brazilian nuclear fuel company
INPRO	International Project on Innovative Nuclear Reactors and Fuel Cycles, an IAEA-inspired international project on developing new nuclear technology
JNFL	Japan Nuclear Fuels Limited, a nuclear fuel services company
KEDO	Korean Peninsula Energy Development Organisation, created to develop the project to build two PWRs in North Korea
KEPCO	Korean Electric Power Corporation, the state-owned major power generation company in Korea

GLOSSARY

KHNP	Korea Hydro and Nuclear Power, the company generating all the nuclear electricity in Korea
KNFC	Korea Nuclear Fuel Company, a fabricator of nuclear fuel
KOPEC	Korean Power Engineering Company, a reactor engineering company and vendor
kWh	Kilowatt hour, a small unit of electricity consumption
KWU	Kraftwerk Union, a German industrial company
LCA	Lifecycle analysis, a method of analysing all the impacts of an energy source from its extraction to final use
LEU	Low enriched uranium, having a uranium-235 assay increased from the natural 0.7% to up to 20%
LLW	Low-level waste, with only minor radioactivity and usually disposed of by incineration or burial
LNG	Liquefied natural gas, commonly conveniently transported in this form
LWR	Light water reactor, both cooled and moderated by normal water
MITI	Ministry of International Trade and Industry in Japan (now supplanted by METI, the Ministry of Economy, Trade and Industry)
MOX	Mixed oxide fuel, a blend of plutonium (from a reprocessing plant) and depleted uranium
MWe	Megawatt electric, a measure of electricity generating capacity
NEA	Nuclear Energy Agency, a constituent organisation of the OECD
NEK	The Bulgarian national electricity company
NIA	Nuclear Industry Association, a lobbying body for nuclear power in the United Kingdom

GLOSSARY

NPT	Treaty on the Non-Proliferation of Nuclear Weapons, the major international agreement curtailing the spread of nuclear weapons and related technology
NSG	Nuclear Suppliers Group, a group of nuclear supplier countries that controls access to nuclear technology and materials
OECD	Organisation for Economic Cooperation and Development, a group of the most developed countries
PBMR	Pebble Bed Modular Reactor, a HTGR being developed in South Africa, to be built in small, modular units
PHWR	Pressurised heavy water reactor, using heavy water (deuterium) as the moderator
PR	Public relations, the activity of presenting a favourable image to the general public
PWR	Pressurised water reactor, one of the two variants of the LWR
R&D	Research and development, the activity of researching new products and then preparing for their introduction to the market
RBMK	A light water cooled and graphite moderated reactor developed in Russia
RepU	Reprocessed uranium, uranium derived from a reprocessing plant
SE	Slovenské elektrárne, the electricity generation company in the Slovak Republic
SNN	Societatea Nationala Nuclearelectrica, the state nuclear power corporation in Romania
SPD	Social Democratic Party, the main left-leaning party in Germany
SWU	Separative Work Unit, a measure of the work done in separating the isotopes in a uranium enrichment facility
TEPCO	Tokyo Electric Power Company, a major company generating nuclear electricity

GLOSSARY

tU	Tonnes of uranium, the internationally accepted measure of uranium mass
TWh	Terawatt hour, a large unit of electricity consumption
VVER	A Russian-designed PWR
WEO	World Energy Outlook, an annual report issued by the IEA
WISE	World Information Service on Energy, an anti-nuclear pressure group
WNA	World Nuclear Association, a trade association of companies involved in nuclear around the world

The World Nuclear University Primer
Nuclear Energy
in the 21st Century

Ian Hore-Lacy

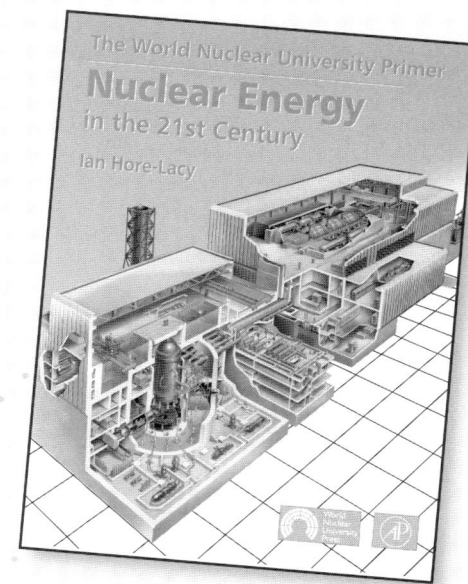

An essential work of reference for specialists and non-specialists alike, providing:

- An introduction to nuclear science
- A comprehensive account of nuclear power today
- Coverage of other nuclear technologies
- Answers to public concerns about safety, proliferation and waste
- Up-to-date data and references

With a Foreword by Dr Patrick Moore, co-founder of Greenpeace

"An invaluable resource for anyone wishing to gain understanding in this crucial field."
- Hans Blix, Director General-Emeritus, International Atomic Energy Agency

World Nuclear University Press